Civil-Military Relations in Postcommunist Europe

Fifteen years after the fall of communism, we are able to appraise the results of the multi-faceted postcommunist transition in Central and Eastern Europe with authority. This volume specifically addresses the complexities of Civil-Military relations throughout this period.

The countries of the region inherited an onerous legacy in this area: their armed forces were part of the communist party-state system and most were oriented towards Cold War missions. They were large in size and supported by high levels of defence spending and were based on universal male conscription. Central and eastern European states have thus faced a three fold civil-military reform challenge: establishing democratic and civilian control over their armed forces; implementing organisational reform to meet the security and foreign policy demands of the new era; and redefining military bases for legitimacy in society.

This volume assesses the experiences of Poland, Hungary, Latvia, Romania, Croatia, Serbia-Montenegro, Ukraine and Russia in these areas. Collectively these countries illustrate the way in which the interaction of broadly similar postcommunist challenges and distinct national contexts have combined to produce a wide variety of different patterns of civil-military relations.

This book was previously published as a special issue of *European Security*.

Timothy Edmunds is a Lecturer in Development and Security at the Department of Politics, University of Bristol

Andrew Cottey is Jean Monnet Chair in European Political Integration and a Lecturer at the Department of Government, University College Cork.

Anthony Forster is Professor of Politics and International Relations and Head of Department at the Department of Politics, University of Bristol.

Civil-Military Relations in Postcommunist Europe

Reviewing the Transition

Edited by
Timothy Edmunds, Andrew Cottey and
Anthony Forster

LONDON AND NEW YORK

First published 2006 by Routledge

2 Park Square, Milton Park, Abingdon, Oxfordshire OX14 4RN
711 Third Avenue, New York, NY 10017

Transferred to Digital Printing 2006

Routledge is an imprint of the Taylor & Francis Group, an informa business

First issued in paperback 2018

Copyright © 2006 Taylor & Francis

All rights reserved. No part of this book may be reprinted or reproduced or utilised in any form or by any electronic, mechanical, or other means, now known or hereafter invented, including photocopying and recording, or in any information storage or retrieval system, without permission in writing from the publishers.

Notice:
Product or corporate names may be trademarks or registered trademarks, and are used only for identification and explanation without intent to infringe.

British Library Cataloguing in Publication Data
A catalogue record for this book is available from the British Library

Library of Congress Cataloging in Publication Data
A catalog record for this book has been requested

ISBN 13: 978-0-415-37631-0 (hbk)
ISBN 13: 978-1-138-37669-4 (pbk)

Contents

Introduction

1 Civil–Military Relations in Postcommunist Europe: Assessing the Transition
Andrew Cottey, Timothy Edmunds & Anthony Forster — 1

Central Europe

2 The Half-Hearted Transformation of the Hungarian Military
Pál Dunay — 17

3 The Transformation of Postcommunist Civil–military Relations in Poland
Paul Latawski — 33

The Baltic States

4 Democracy and Defence in Latvia: Thirteen Years of Development: 1991–2004
Jan Arveds Trapans — 51

South Eastern Europe

5 Civil–military Relations in Croatia: Politicisation and Politics of Reform
Alex J. Bellamy & Timothy Edmunds — 71

6 The Transformation of Romanian Civil–military Relations: Enabling Force Projection
Larry L. Watts — 95

7 Civil–military Relations in Serbia–Montenegro: An Army in Search of a State
Timothy Edmunds — 115

The Former Soviet Union

8 Vladimir Putin and Military Reform in Russia
Dale R. Herspring — 137

9 Ukraine: Reform in the Context of Flawed Democracy and Geopolitical Anxiety
James Sherr — 157

Index — 175

Civil–Military Relations in Postcommunist Europe: Assessing the Transition[1]

ANDREW COTTEY,* TIMOTHY EDMUNDS** & ANTHONY FORSTER**
*Department of Government, University College Cork, Cork, Republic of Ireland, **Department of Politics, University of Bristol, Bristol, UK

Fifteen years after the fall of communism and the end of the Cold War provides sufficient time to make an assessment of the multi-faceted postcommunist transition in Central and Eastern Europe. This special issue of *European Security* attempts such an assessment in the area of civil–military relations. When communism collapsed in 1989–91 the countries of Central and Eastern Europe inherited a very particular legacy in relation to armed forces, defence policy and civil–military relations: the armed forces were part of the communist party-state system; they were oriented, except in the cases of Yugoslavia, Albania and arguably Romania, towards the Cold War mission of conflict with the West; they were large in size and supported by high levels of defence spending; and they were based on universal male conscription, which gave them a broad social presence and perhaps made them a unifying social force. Anyone who travelled in communist Eastern Europe or the Soviet Union from the late 1940s to the late 1980s witnessed highly militarised societies, with uniformed soldiers a universal presence.

The countries of Central and Eastern Europe therefore faced broadly similar challenges: reforming the communist party-state system of civil–military relations and replacing it with, hopefully, democratic models of civil–military relations; reducing the size of the armed forces and defence spending and reorienting the military towards new post-Cold War missions; and building new bases of military–society relations. Drawing on the broader civil–military relations literature this special issue assesses the transition in civil–military relations, focusing on these three areas: democracy and the military; defence reform and professionalisation; and the military and society. The articles in this volume examine the postcommunist civil–military transition in eight different countries: Poland, Hungary, Latvia, Romania, Croatia, the Federal Republic of Yugoslavia (FRS)/Serbia–Montenegro,[2] Ukraine and Russia. Collectively these countries illustrate the way in which the interaction of broadly similar postcommunist challenges and distinct national contexts (in terms of size, historical legacies, geo-strategic location, political and economic transitions, involvement in violent conflict and integration with NATO and the European Union) have combined to produce a variety of different patterns of civil–military relations. This introduction draws overall conclusions on the transition in civil–military relations in postcommunist Europe. It suggests that the relative, although never absolute, homogeneity of civil–military relations in communist Europe is being replaced by greater diversity: those countries in the process of joining NATO and the EU are developing patterns of civil–military relations similar to those in the long-established democracies of Western Europe and North America; civil–military relations in the former Yugoslavia republics, especially Serbia–Montenegro, have been deeply scarred by the wars of the 1990s and thus face particular problems relating to armed forces roles in those wars, including issues of responsibility for war crimes; Russia, Ukraine and the other former Soviet republics appear to be moving towards situations where civil–military relations are one part of semi- or 'soft' authoritarian regimes, while economic problems have resulted in a more general degrading—de-professionalisation—of the military.

Democracy and the Military

Military intervention in domestic politics, and the degree of political independence and influence of the military, are generally seen as one of the key problems, often *the* central challenge, of civil–military relations. Much of the academic civil–military relations literature is thus devoted to exploring the circumstances that give rise to and the factors that explain military intervention in politics. Much attention is also paid to exploring circumstances and factors that facilitate civilian political control of the military and the establishment of democratic civil–military relations.[3] The communist system as it developed in

Eastern Europe and the Soviet Union resulted in the emergence of a very particular relationship between the military and politics. The military was generally under the strict political, but of course not democratic, control of the civilian communist leadership and had quite limited room for independent political action. At the same time, the military was politicised in the sense that it was one of the vehicle's for society-wide inculcation of communist values. In most cases it was intertwined with the communist party through the establishment of party cells and the oversight of party education officers throughout the armed forces. Finally, although the military's political independence was limited, the military high command retained substantial control over defence policy, military strategy and force structure.[4] As communism crumbled in the late 1980s and early 1990s this civil–military context raised major questions about the role of the military. To what extent was the military loyal to communism and to what extent would it, either independently or at the behest of or in conjunction with civilian communist leaders, act to defend the old regime? Once the communist regimes fell, the question shifted: to what extent would the armed forces support or resist a transition to *democratic* civilian political control and the broader de-communisation of the military?

Prior to and during the 1989 revolutions in Eastern Europe there were fears that the military might intervene forcefully to defend the communist regimes. In the event this did not happen, suggesting that, despite forty years of communist penetration, military loyalty to communism in the non-Soviet Warsaw Pact states was probably skin deep. Lech Walesa, former leader of Poland's anti-communist Solidarity movement and the country's first postcommunist President, argued that the Polish military was like a radish, red (communist) on the outside but white (national) on the inside.[5] As the articles on Poland, Hungary and Romania in this volume indicate the relative ease with which these countries have consolidated democratic civilian control of their armed forces, and the absence of significant opposition to this process from the military itself, reinforces the argument that military loyalty to communism was always skin deep in these states. The situation in the Soviet Union and the Socialist Federal Republic of Yugoslavia (SFRJ) was different. In the Soviet Union elements of the military high command were wary of the reforms introduced by President Mikhail Gorbachev in the late 1980s and played a central role in the August 1991 coup against Gorbachev—indicating the greater loyalty of at least some parts of the Soviet military to communism compared to their Eastern European counterparts. Russian President Boris Yeltsin's success in opposing the coup and the subsequent disintegration of the Soviet Union, however, suggested that the military was both divided and ill prepared to actually implement a coup.

In the SFRJ the Yugoslav National Army (JNA) became closely intertwined with the politics and wars that emerged at the beginning of the 1990s. The JNA was one of the key national institutions that had helped to hold the ethnically fragmented SFRJ and its own institutional identity was closely linked to that of the Federal state. This placed it in natural opposition to the secessionist claims of Slovenia and Croatia and ultimately led to its cooption by the Milošević regime and its Serbian nationalist project. The JNA became an increasingly Serbian institution, eventually becoming the army of the successor FRY (as the Yugoslav Army or VJ) and now today's Serbia–Montenegro (as the Armed Forces of Serbia–Montenegro or VSCG). The military always retained a strong tradition of professionalism and corporate self-governance, which made it resistant to the more aggressive politicisation attempts of the Milošević regime. Nevertheless, by the end of the 1990s, its leadership had been filled with Milošević cronies and as an institution it was inextricably linked to the regime's nationalist project and wars. A decade later the challenges of civil–military relations in Serbia–Montenegro are part of the larger challenge of dismantling that nationalist project and the other pathologies, especially corruption and political cynicism, that it imposed on the country. More widely the future of civil–military relations in the country will be closely connected to the future viability of the state union itself, uncertainty over which has stalled defence reform and intensified domestic political divisions. As Alex Bellamy and Timothy Edmunds discuss in this volume, nationalism and war resulted in similar attempts by the Tuctman regime to politicise the Croatian armed forces and tie them to the Croatian nationalist project. Zagreb's road away from these problems is proving easier than Belgrade's however. This is because, in contrast to Serbia–Montenegro, the Croatian state represents a finished political project. The period since 2000 has also seen the emergence of an increasingly strong domestic political consensus on the direction of the country's foreign and defence policy which has allowed Croatia to press ahead with it civil–military reform process.

Following the collapse of communism, the new democratic regimes in Central and Eastern Europe faced the challenge of establishing *democratic* civilian political control of the military and reforming the institutions for the management of the armed forces. As the papers by Paul Latawski, Pal Dunay, Larry Watts and Jan Trapans in this volume illustrate, this challenge proved easier than might have been expected. A core group of Central and Eastern European states—Poland, the Czech Republic, Slovakia, Hungary, Romania, Bulgaria, Estonia, Latvia, Lithuania and Slovenia—have all made substantial progress in entrenching democratic civilian control of the military. In all these states major reforms were undertaken in the early and mid-1990s. New constitutions establishing the principle of democratically controlled militaries were put in place, new chains of command were instituted to operationalise this principle, and the military were depoliticised as the old link between the armed forces and the communist party was broken. Although there were often disputes over the details of such reforms and the old guard in the senior military leadership sometimes obstructed their implementation, there was almost no serious opposition to the principle of civilian democratic control of the military. Where minor crises did occur, moreover, they usually involved disputes between civilian politicians (for example, between Presidents and government ministers) over should control the military, rather than the potentially more serious problem of the military asserting its own right to have a say in politics. There is a strong case that in the sphere of civil–military relations, as more generally, this group of Central and Eastern European states have now reached the stage of democratic consolidation: the principle of democratic civilian control of the military is deeply entrenched and not seriously challenged; institutions have been put in place to put this principle into practice and these institutions function reasonably effectively; and although politics in these countries is sometimes messy and chaotic, serious threats to democracy from the civil–military sphere appear unlikely.

The consolidation of democratic civil–military relations in Central and Eastern Europe also raises interesting questions about how far there are different models of democratic civil–military relations in the region. It is striking that all these countries have adopted models of democratic civil–military relations in which the military is implicitly an institution separate from the rest of society and democracy is exercised in essence through the control of the military by the democratically elected President, government and parliament. This contrasts with the post-1945 Federal Republic of Germany's model, and also the longer-standing Swiss model, of a closely integrated military and society, in which the active participation of civilians in the armed forces through conscription is a central element of ensuring the democratic character of the military.

While the Central and Eastern European states moved down the road towards democratic consolidation in the 1990s, Russia, Ukraine and the other former Soviet republics have faced more difficult political transitions, especially from a democratic perspective. In some cases—as in Belarus and the Central Asian states—communist authoritarianism has been replaced by an old-new brand of postcommunist authoritarianism, with former republican communist party chiefs establishing themselves as national leaders and much of the old communist era political infrastructure remaining in place. In Russia and Ukraine, a more complex situation has emerged with elected Presidents and parliaments in place, a variety of political forces operating but political power increasingly concentrated in the hands of the President and their supporters and weak counterbalancing institutions. Critics thus warn that Vladimir Putin's Russia and Leonid Kuchma's Ukraine have become, or risk becoming, semi- or 'soft' authoritarian states in which the concentration of power in the hands of the President and their circle fundamentally undermines democracy. In this context, the military and other security forces such as the interior ministry and intelligence services are some of the key instruments of presidential power.

While much of the academic literature on civil–military relations focuses on the issue of military intervention in domestic politics, one of the main challenges facing the countries of postcommunist Europe has been the related but distinct problem of establishing effective control over defence policy. As we noted above, one feature of communist civil–military relations was the relatively high degree of autonomy given to the military in relation to defence policy. As a consequence, when communism collapsed almost no structures existed for civilian democratic control of defence policy. The new governments in Central and Eastern Europe faced a daunting array of challenges in this area: appointing civilian defence ministers and civilian staff to previously military dominated defence ministries; separating military general staffs from defence ministries and subordinating the former to the latter; putting in place institutions for the management of defence policy, budgets and procurement; and establishing parliamentary committees and procedures for the oversight of defence policy. These issues can be viewed as a set of second generation challenges following on from the basic first generation challenge of ensuring that the military does not intervene in domestic politics.[6] For much of postcommunist Europe these second generation problems have proved far more challenging than the first generation ones and have comprised the real agenda for civil–military relations since the early 1990s.

As the articles on Poland, Hungary Romania and Latvia in this volume show, the countries of Central and Eastern Europe spent much of the 1990s introducing a series of reforms designed to address these second generation problems.[7] These reforms proved problematic for two reasons: they involved complex institutional and technical issues, but they also threatened the previous power of the military to shape defence policy and were therefore quite often resisted by senior officers. Despite this, the core group of Central and Eastern European states identified above made substantial progress in implementing such reforms in the 1990s and now mostly have reasonably functioning systems for democratic civilian control of defence policy. Similar patterns have been visible in Croatia from 2000 onwards. This is not to say that these countries do not face problems in these areas. There continue to be problems of mismatches between declared defence policy goals and the resources available to meet these goals, inadequate transparency in relation to defence budgets and procurement and periodic scandals of issues such as procurement and corruption. The longer-standing democracies of Western Europe and North America, however, are hardly immune to these problems themselves and their existence in the newer democracies of Central and Eastern Europe does not by and large reflect any wider malaise in civil–military relations. In the former Soviet republics, much less progress has been made in implementing these types of reforms. More than a decade after the end of communism while Russia and Ukraine both have civilian defence ministers, the boundaries between defence ministry and general staff remain blurred, their defence ministries remain predominantly military-staffed organisations, defence policy-making, budgeting and procurement are deeply opaque and parliamentary oversight is limited at best. These problems are part of and reinforce a deeper malaise illustrated by James Sherr and Dale Herspring's contributions to this volume. Corruption and incompetence have become endemic in the armed forces of Russia, Ukraine and the other former Soviet republics. Thus, whereas the civil–military problematique has traditionally been conceived in terms of the relationship between the military and the political sphere and the danger of military praetorianism, in the former Soviet Union (and also parts of the former Yugoslavia) the problem is now the impact of wider economic and social problems on the military and the danger that these problems may undermine the military's ability to operate as an effective and useful state institution. Political leaders in these countries are struggling with these problems but, as Dale Herspring's discussion of Russian President Vladimir Putin's approach to military reform shows, whether they can find effective solutions is open to question.

Defence Reform and Professionalisation

The countries of Central and Eastern Europe and the former Soviet Union inherited large, conscript-based armed forces structured for Cold War military missions. In the new political and strategic environment that emerged in the 1990s it was clear that the countries of postcommunist Europe had to radically reform their defence policies and armed forces, but quite what this should mean in terms military strategy, defence spending, force structure and procurement was unclear. In the early 1990s, driven by the end of the Cold War, economic realities and a more general desire to demilitarise their societies, the immediate response of most governments in the region was to drastically reduce defence expenditure and cut the overall numbers in the armed forces. These initial reforms left open the longer-term questions of what these countries' armed forces were actually for and what this implied for defence policy.

The most obvious drivers of defence policy are external military threats to national security. For most Central and Eastern European states Russia was—rightly or wrongly—perceived to be the most likely aggressor. This logic drove these states' desire to join NATO. Other neighbouring Eastern European states were to some extent also perceived as potential threats, as in the historically antagonistic relations between Hungary and its neighbours or Poland and its neighbours, but in most cases such contingencies were seen as less serious than the potential threat posed by Russia. However, while the Central and Eastern European states sought membership of NATO and did reorient their forces away from Cold War deployments on their western borders, concern about Russia did not result in a dramatic redeployment of forces towards their eastern borders or significant increases in defence spending. Implicitly, Russia was viewed more as a potential long-term threat than an immediate military danger. The reality for most of the states of postcommunist Europe was that with the end of the Cold War they had entered a 'low threat' era, at least in terms of classical state-based military threats to their national security. During the 1990s the debate on the role of armed forces shifted away from issues of defending national territory in much of postcommunist Europe and towards the question of how these countries might contribute to international security more generally. Not surprisingly, the Yugoslav conflict was the primary driver of this process. Central and Eastern European leaders feared that the conflict in the former Yugoslav might destabilise the region more broadly, while Western governments encouraged Central and Eastern European states to contribute to international efforts to manage the conflict. As a consequence, after NATO's interventions in Bosnia in 1995 and Kosovo in 1999, Central and Eastern European states contributed forces to the NATO-led follow-on peacekeeping forces. The ability to deploy peacekeeping forces, and the interrelated desire to integrate with NATO became a — to some extent *the* — key driver of military reform. The September 2001 terrorist attack reinforced but also altered this dynamic. The ability to contribute forces to counter-terrorist operations and related intervention and peacekeeping missions became a further driver of defence policy. A number of Central and Eastern European states have contributed forces to the post-war stabilisation missions in Afghanistan after the fall of the Taliban and in Iraq after the fall of Saddam Hussein. The addition of the counter-terrorist mission, however, also implied that armed forces might need the capability to conduct more robust war fighting and peace enforcement missions as well as 'softer' peacekeeping tasks.

These various drivers of military reform have resulted in different patterns across the region. For a small group of countries, direct involvement in postcommunist conflicts became the central driver of military reform. Serbia–Montenegro, Croatia and Bosnia-Hercegovina (BiH — or to be more accurate the various statelets and groups that constituted BiH) developed armed forces to fight in the wars of Yugoslav dissolution, with spending and force structures reflecting this. Further afield, a similar dynamic was discernible between Armenia and Azerbaijan in relation to the Nagorno–Karabakh conflict. For the core group of Central and Eastern European states noted above, however, the ability to contribute forces to peacekeeping and intervention missions beyond their national borders drove military reforms in the 1990s. This resulted in the development of small elite forces and units for deployment on peacekeeping missions, combined with a focus on inter-operability with NATO (for example, in terms of communications equipment and command and control procedures). These reforms were successful in the sense that they enabled these states both to contribute to peacekeeping missions and to integrate their forces with those of the longer-standing NATO members. Critics, however, charged that the Central and Eastern European states risked developing two-tier armed forces, with a small elite of military forces capable of working with NATO and a large rump of increasingly downgraded armed forces incapable of providing for a credible national defence should that be needed.[8] In Russia and Ukraine a very different set of dynamics have been at work. Although both countries have contributed forces to peacekeeping missions, this has not been a central driver of defence policy. In both countries economic realities have in many ways been the central shapers of defence policy: dramatic economic collapse resulted in an equally dramatic collapse in the funds available for the armed forces. Combined with the break-up of the Soviet Union, this has resulted in a significant downsizing of the armed forces, along with major cuts in training and related activity and an effective freeze on new procurement. Dale Herspring has described this process as not military reform in any meaningful sense but rather the de-professionalisation of the Russian armed forces.[9]

These debates on defence reform relate to the concepts of professional armed forces and professionalisation. Some of the academic literature on civil–military relations has placed particular emphasis on professionalisation, arguing that this is a central determinant of the prospects for civilian control of the military since professional armed forces are more likely to focus on their professional military tasks of preparing for national defence and the like, rather than intervening in domestic politics.[10] This debate, however, creates some confusion, since the term professional armed forces is open to different interpretations. In this context, professional may refer to the extent to which the armed forces focus on their professional military tasks (as distinct from intervening in domestic politics), the degree to which the military is professionally competent (i.e., capable of performing their military functions effectively) or the distinction between an all-volunteer military and a conscript-based force. In postcommunist Europe it is the latter definition—i.e., equating a professional military with an all-volunteer force—that has come to the fore. Although the process has developed slowly since the 1990s, those Central and Eastern European states who are joining or aspire to join NATO and the EU appear to be gradually moving towards either predominantly or completely volunteer forces. Periods of conscription have been reduced and a number of states have now made commitments to move towards all-volunteer forces.[11] This process has in significant part been driven by the requirement to provide troops for peacekeeping missions and capable of operating alongside other NATO forces. It has, however, tended to preclude wider debate on the defence policy choices facing Central and Eastern European states. From a national defence perspective, for example, there may be case for retaining territorial defence forces and conscription alongside the development of 'professional' forces for deployment beyond national borders. At a minimum, there should be scope for a broader debate on the range of defence policy choices open to the states of postcommunist Europe and the longer-term implications of different paths.

The Military and Society

Communism resulted in the emergence of a very particular yet contradictory relationship between the military and society. The military was one of the tools of state power used to maintain the communist regimes and was therefore sometimes an unpopular institution, especially in Eastern Europe and parts of the Soviet Union, such as the Baltic republics, where anti-Soviet nationalism remained strong. At the same time, the military was sometimes seen as independent national institution (in Romania most obviously, but also for example in Poland). Conscription, further, was arguably a social unifying force, while the military was used, particularly in the Soviet Union, as a means of nation-building and social engineering. We are right to be sceptical about the extent to which communism created a new 'socialist man'. However, the strength of nostalgia for the Soviet era in much of the former Soviet Union more than a decade after the fall of communism and the extent to which the military is still viewed positively because of its role in the Great Patriotic War against German fascism suggest that national identities, including their military dimensions, *were* deeply shaped by communism.

The fall of communism and the end of the Cold War meant that the old roles of the military—defenders of the communist regimes against domestic threats, Cold War defenders against the threat from the West and a force for communist nation-building—were swept away overnight. Some of the changes discussed above—reductions in defence spending, the size of armed forces and conscription periods—resulted in a general demilitarisation of society. This process has progressed furthest in those countries closest to the West (both politically and geographically), but less dramatically in the former Soviet Union and the former Yugoslavia. What new patterns of military–society relations might replace the old communist model was far less clear.

One of the longest-standing social roles of armed forces, in a wide variety of different historical and political circumstances as has been as nation-builders: both directly as the means by which a nation may be united by force or achieve independence from foreign oppressors and indirectly as a means of promoting national unity, in particular through conscription. Given the more general rebirth of nations in postcommunist Europe one might have expected to have seen a wider re-emergence of the idea of armed forces as nation-builders. As Alex Bellamy and Timothy Edmunds discuss in their contribution to this volume, this did occur in the context of the so-called Homeland War in Croatia in the early 1990s. In general, however, armed forces have not been seen or used as central tools of nation-building in postcommunist Europe. In countries such as Poland, Hungary and Romania this may reflect the fact that these states were already quite well established and in some ways beyond the stage of nation-building. In other countries, such as Ukraine, the very weakness of nationalism may have militated against using the military as a central tool of nation-building. In FRJ and subsequently Serbia–Montenegro, the military's nation-building role has been hampered by continuing uncertainties and disagreements over the future of the state itself.

The demise of the military's communist regime defence role, the limited extent to which the military has become a force for nation-building in postcommunist Europe and the generally low-level of direct external threat to national security raised major questions about role the military would play and what this meant for its relations with wider society. As we discussed above, one major response to this dilemma has been increasing involvement in peacekeeping and intervention operations beyond national borders. Another element of this has been the use of the military as a means of promoting bilateral and multilateral cooperation with neighbouring states, for example through military exchanges and joint exercises. In terms of military–society relations we describe this process as the armed forces taking on a military diplomacy role. In effect the military becomes an external, foreign policy expression of society's broad goals and values. In much of the postcommunist region therefore the trend towards participation in peacekeeping operations reflects a desire on the part of governments and populations to be seen as a good international citizens, contributing to wider regional and international security and the promotion of democratic values. Although there have been controversies, in particular over NATO's 1999 intervention in Kosovo and the 2003 Iraq War, public opinion has generally been supportive of participation in such operations and willing to accept at least small numbers of casualties in support of this cause.

Another striking feature of military–society relations in postcommunist Europe has been the re-emergence of patterns reflecting pre-communist historical legacies. Countries such as Poland and Romania, for example, have relatively martial histories, in which the military has usually been viewed by society at large as an important and positive national actor. In contrast, countries such as the Czech Republic and Hungary have less martial histories, in which the military has often been viewed as an ineffective defender of national sovereignty and a relatively less important national institution. These patterns have re-emerged in these countries since the fall of communism, with the military generally viewed very positively in Poland and Romania but less so in the Czech Republic and Hungary. Given that these countries have only recently emerged from four decades of communism, this suggests that broad societal attitudes to the military may be remarkably deeply embedded and enduring, passed on from generation to generation.

The changing patterns of military–society relations in postcommunist Europe also raises questions about how far such changes reflect more general global trends. Charles Moskos, John Allen Williams and David Segal have developed the concept of the 'postmodern military' to describe a variety of changes underway in North American, Western European and arguably other armed forces.[12] Postmodern militaries are defined by the combination of a shift away from defence of national territory and towards peacekeeping and intervention operations; associated shifts towards all-volunteer armed forces and increasingly sophisticated military technology; a breakdown of traditional military hierarchies and military–civilian boundaries; the demise of traditional deference within society, including towards institutions such as the military; and the way in which broader social issues—such as women's and gay rights —impinge on the military. Some of these trends—such as the move towards peacekeeping and all or predominantly volunteer forces—are clearly observable in postcommunist Europe. Other trends, for example a broad societal lack of interest in armed forces in much of the region, may also fit the postmodern model. Yet others, such as issues of women's and gay rights within the armed forces, have barely moved onto the agenda. This suggests that while armed forces in postcommunist Europe may gradually be moving towards the postmodern model observable in the longer-established democracies, this is likely to be a slow and gradual process rather than a dramatic transition.

The Drivers of Change

What factors have been driving the changes in civil–military relations in postcommunist Europe and what does this tell us about the likely future prospects for civil–military relations in the region and more generally? Most obviously, the changes in civil–military relations in Central and Eastern Europe and the former Soviet Union are the result of the collapse of communism a decade and a half ago. This points to the obvious but sometimes neglected conclusion that the broad domestic political environment is the key factor shaping civil–military relations. Thus, it is no coincidence that those countries that that have made most progress in democratization in general have also made most progress in democratizing civil–military relations. Similarly, civil–military relations in Russia, Ukraine and the other former Soviet republics reflect the broader pattern of partial democratization combined with strong and sometimes authoritarian presidential rule.

A second set of factors relates to the impact of external developments on civil–military relations. From a theoretical perspective, analysts have explored the impact of the extent of external security threats faced by a state on the likelihood of military intervention in domestic politics.[13] The relatively low level of external threat faced by most Central and Eastern European states and the relative ease with which they have consolidated democratic civilian control of the military supports the argument that a low level of external threat reduces the power and importance of the military and thereby enhances the prospects for democratic civilian control of armed forces. This argument is reinforced by the fact that the few cases in postcommunist Europe where the military have been more directly politicised—in particular Croatia under Tudman and FRY under Milošević—have taken place in a context of significantly higher levels of external threat or conflict. A second external factor, much less discussed in the general civil–military relations literature, is the existence of positive external pull factors shaping civil–military relations. In Central and Eastern Europe, NATO and the EU—and the existence of a broader community of wealthy established democracies in the West—have exerted a very powerful influence on civil–military relations. At a strategic level, NATO/the EU/the West have provided a model to aspire to, an implicit bastion of support for democratic reformers and hard leverage to support reforms (in particular through the implicit conditionality attached to membership of NATO and the EU). Functionally, NATO, the EU and individual Western governments have provided practical support for reforms and frameworks for the multilateralisation and reinforcement of the civil–military reform process. Integration into NATO in particular but the West more generally, has become a key driver of defence policy in much of Central and Eastern Europe.

A third set of factors often discussed in the wider literature is the impact of changes in military technology and strategy on civil–military relations. The emergence of the modern nation-state was intimately associated with the development of mass armies and conscription. The development of armoured warfare, modern airpower and nuclear weapons similarly re-shaped civil–military relations, resulting in the development of more technically sophisticated professional soldiers. As was noted above, the combination of low levels of external threat, a shift towards peacekeeping and intervention operations and the so-called Revolution in Military Affairs are pushing the Central and Eastern European states towards pre-dominantly or all-volunteer armed forces. This in turn impinges on other aspects of civil–military relations, in particular reducing the military's role as a force for nation-building or social unification and more generally further demilitarising society. The long-term implications – and indeed viability – of armed forces based on a small professional military designed for operations beyond national borders and with limited engagement with wider society remains to be seen.

A final set of factors relates to the impact of broader social changes on civil–military relations. The postmodern military concept is based in part on the argument that, at least in the industrialised West, we are witnessing the emergence of postmodern societies different in important ways from their nineteenth and twentieth century modern, industrial predecessors. The extent to which the countries of postcommunist Europe and Eurasia are moving down this path remains open to debate. Since the collapse of communism these countries have experienced the collapse of old political authority structures and neo-liberal economic reforms and the associated retreat of the state, thereby introducing some of the elements of postmodern societies seen in the West. At the same time, however, much of postcommunist Europe and Eurasia arguably retain quite traditional, conservative beliefs in relation to issues such as nation, religion, gender and family. As societies there is a case therefore that the countries of postcommunist Europe are currently torn between the postmodern and the earlier modern (and sometimes even pre-modern) eras. In a broad sense civil–military relations in the region are likely to reflect this.

Conclusion

A decade and a half after the fall of communism what conclusions can be drawn about the state of civil–military relations in postcommunist Europe and the nature of the region's transition in civil–military relations? The relative homogeneity of communist civil–military relations has been replaced by significant diversity across the region. The Central and Eastern European states that have joined NATO and the EU (and more recently NATO aspirants such as Croatia) have consolidated democratic civilian control of their militaries, reoriented their defence policies towards peacekeeping and intervention operations beyond their borders and are developing postmodern military–society relations where the military is simply one state institution alongside others. In contrast, in Russia, Ukraine and most of the other former Soviet republics the military has become part of the nexus of semi- or outright authoritarian presidential rule, while severe economic and social problems are resulting in a dramatic downgrading of the military's professional and operational competence and severely inhibiting the prospects for meaningful military reform. In the former Yugoslavia, in particular in Serbia under Milošević and Croatia under Tudman, war brought the military more directly into domestic politics, but again largely as a result of authoritarian presidential rule. The Yugoslav wars also made war crimes, corruption and links to organised crime central elements of civil–military relations.

The end of communism triggered the emergence of an academic sub-discipline analysing the postcommunist transition process—the so-called 'transitology'. This raises broader questions about the extent to which countries are still in an on-going postcommunist transition process or have now reached a new situation that can be said to have stabilised. As was argued above, the Central and Eastern European states that have joined NATO and the EU have probably passed the point of democratic consolidation in civil–military relations. It is perhaps more useful therefore to analyse them as democracies like their Western European counterparts struggling with the on-going problems of adjusting armed forces to a post-Cold War and post-9/11 world (albeit facing some particular constraints, especially in the economic sphere), rather than states still overcoming the legacy of communism in civil–military relations. The presidential (semi-) authoritarianism that has emerged in much of the former Soviet Union represents a very different pattern with its own distinctive civil–military dynamics. The extent to which this pattern is stable is a matter for debate. Across much of the former Soviet Union Presidents have consolidated their hold on power and opposition is generally weak and/or divided. Presidential elections, the death of old rulers and/or the hand over of power to new ones are, however, likely to present periodic challenges to the status quo. In these circumstances, control of military power may play an important role and the military may be drawn into domestic politics (as occurred in the 1993 Russian parliamentary coup). The recent 'rose revolution' in Georgia illustrated the potential for dramatic political change and democratic opposition to presidential strong-men. In the former Yugoslavia, just as war and authoritarianism brought the military into politics, peace and political change have raised major questions about the military's role in the new environment. In Croatia this resulted in a more general demilitarisation of society from the mid-1990s and a rapid democratisation of civil–military relations from 2000 onwards. In Serbia–Montenegro the post-war transition has proved much more problematic and attempts to reform civil–military relations have been severely constrained by a more general crisis in domestic politics and the legitimacy of the state itself. In the former Soviet Union and much of the former Yugoslavia therefore the transition in civil–military relations is far from complete.

This analysis of civil–military relations in postcommunist Europe also has implications for the study of civil–military relations more generally. First, the relative absence of military intervention in domestic politics in postcommunist Europe as a whole and the consolidation of democratic civilian control of the military in those countries that have joined NATO and the EU also reflects a broader global trend away from military praetorianism.[14] There is scope for further analysis of the how the global democratic revolution of the last few decades—the spread of democracy to an increasing number of countries around the world—is affecting civil–military relations and the nature of civil–military relations in a post-praetorian world. Second, given the past salience of the military as a symbol of the nation and a tool of nation-building, it is remarkable that, with some exceptions such as Croatia, armed forces have not played a more prominent nation-symbolising and nation-building role in postcommunist Europe. This trend is being further reinforced by the shift away from conscription and towards volunteer armed forces. The issue of the role of armed forces and military–society relations in a post-national era also needs more analysis. Third, developments in postcommunist Europe also point towards the importance of the political-economy of civil–military relations. While most analyses of civil–military relations have tended to focus on either the military's role in politics or broader military–society relations, some of the central problems in civil–military relations in postcommunist Europe have related to economic issues, in particular the impact of dramatic cuts in defence spending on the military, corruption within the armed forces and links between the armed forces and organised crime. Similar problems have occurred in other countries and regions as a consequence of post-authoritarian transitions, wars and post-war situations and externally driven neo-liberal economic policies.[15] Yet there has been very little sustained analysis of the role of military and security forces in these new economic dynamics. In postcommunist Europe and elsewhere there is certainly a need for more serious analysis of these troubling new economic dimensions of civil–military relations.

Notes

[1] This special issue of *European Security* draws together conclusions and analysis from a research project on 'The Transformation of Civil–military Relations in Comparative Context' funded by the UK Economic and Social Research Council (ESRC) 'One Europe or Several?' programme (award number L213 25 2009). Further studies from the project can be found in Andrew Cottey, Anthony Forster & Timothy Edmunds (eds), *Democratic Control of the Military in Postcommu-*

nist Europe: Guarding the Guards (Houndmills: Palgrave-Macmillan, 2002), Anthony Forster, Timothy Edmunds & Andrew Cottey (eds), *The Challenge of Military Reform in Postcommunist Europe: Building Professional Armed Forces*, (Houndmills: Palgrave Macmillan, 2002) and Anthony Forster, Timothy Edmunds & Andrew Cottey (eds), *Soldiers and Societies in Postcommunist Europe: Legitimacy and Change* (Houndmills: Palgrave Macmillan, 2003). The project website can be accessed at http://civil–military.dsd.kcl.ac.uk/

[2] The Federal Republic of Yugoslavia became the confederative State Union of Serbia and Montenegro in February 2003.

[3] The classical works on this subject are Samuel P. Huntington, *The Soldier and the State: The Theory and Practice of Civil–military Relations* (Cambridge, MA: The Belknap Press of Harvard University, 1957); Morris Janowitz, *The Professional Soldier: A Social and Political Portrait* (London: Collier-Macmillan, 1960); and Samuel E. Finer, *The Man on Horseback: The Role of the Military in Politics* (London and Dunmow: Pall Mall Press, 1962). For more recent contributions see Martin Edmonds, *The Armed Services and Society* (Leicester: Leicester University Press, 1988); Larry Diamond & Marc F. Plattner (eds), *Civil–military Relations and Democrac* (Baltimore, MD: The Johns Hopkins University Press, 1997) and Michael Desch, *Civilian Control of the Military: The Changing Security Environment*, (Baltimore, MD: The Johns Hopkins University Press, 1999).

[4] On communist civil–military relations see Roman Kolkowicz, *The Soviet Military and the Communist Party* (Princeton NJ: Princeton University Press, 1967); Dale R. Herspring & Ivan Volgyes (eds), *Civil–military Relations in Communist Systems*, (Boulder, CO: Westview Press, 1978); Timothy J. Colton, *Commissars, Commanders and Civilian Authority: The Structure of Soviet Military Politics* (Cambridge, MA and London: Harvard University Press, 1979).

[5] Thomas S. Szayna, *The Military in a Postcommunist Poland*, N-2209-USDP (Stan Monica, CA: The RAND Corporation, 1991), p. 43.

[6] We have developed this argument in Andrew Cottey, Timothy Edmunds & Anthony Forster, 'The Second Generation Problematic: Rethinking Democracy and Civil–military Relations', *Armed Forces & Society*, 29:1 (Fall 2002), pp. 31–56.

[7] For further detailed analyses of these types of reforms, and the problems encountered in implementing them see Jeffrey Simon's recent works *Hungary and NATO: Problems in Civil–military Relations* (Lanham, MA: Rowman & Littlefield, 2003), *Poland and NATO: A Study in Civil–military Relations* (Lanham, MA: Rowman & Littlefield, 2004) and *NATO and the Czech and Slovak Republics: A Comparative Study in Civil–military Relations* (Lanham, MA: Rowman & Littlefield, 2004).

[8] Christopher Donnelly, 'Reshaping Armed Forces for the 21st Century', *NATO Think Piece*, 10 August 2001, http://www.nato.int/docu/articles/2000/a000913a.htm (accessed 15 March 2005; Timothy Edmunds, 'NATO's New Members', *Survival*, 45: 3 (Autumn 2003).

[9] Dale R. Herspring, 'De-professionalising the Russian Armed Forces', in Anthony Forster, Timothy Edmunds & Andrew Cottey (eds), *The Challenge of Military Reform in Postcommunist Europe: Building Professional Armed Forces*, (Houndmills: Palgrave Macmillan, 2002), pp. 197–210.

[10] Huntington, *The Soldier and the State*.

[11] Anthony Forster, Timothy Edmunds & Andrew Cottey (eds), *The Challenge of Military Reform in Postcommunist Europe: Building Professional Armed Forces*, (Houndmills: Palgrave Macmillan, 2002).

[12] Charles Moskos, John Allen Williams & David R. Segal, *The Postmodern Military: Armed Forces After the Cold War* (Oxford: Oxford University Press, 2000).

[13] Desch, *Civilian Control of the Military*.

[14] See for example, Robin Luckham, Democratic Strategies for Security in Transition and Conflict', in Gavin Cawthra & Robin Luckham (eds), *Governing Insecurity: Democratic Control of Military and Security Establishments in Transitional Democracies* (London: Zed Books, 2003).

[15] Mats Berdal & David Malone (eds), *Greed and Grievance: Economic Agendas in Civil Wars* (Ottawa: International Development Research Centre, 2000).

The Half-Hearted Transformation of the Hungarian Military

PÁL DUNAY
Senior Researcher, Stockholm International Peace Research Institute (SIPRI), Solna, Sweden

When historians look back to the transformation of Central and Eastern Europe after the Cold War they will most probably conclude that the transition of Hungary was among the most successful. After severe economic decline the country recovered and by the end of the late 1990s the per capita GDP exceeded that of the late 1980s. Democratic institutions were established and have functioned properly in most cases. The support of the population for liberal democracy has increased and has no alternative. This irreversible development has been recognised by the West and its institutions. Between 1990 and 2004 Hungary joined a number of Euroatlantic institutions, ranging from the Council of Europe (1990), the OECD (1996), NATO (1999) and the European Union (EU) (2004). The grass root integration of the country has brought even more results than formal membership. Since the beginning of 2000 the EU has accounted for approximately three-quarters of Hungary's total exports. This is higher than the share of intra-EU exports of 13 of the 15 member-states which belonged to the EU before May 2004.[1]

In spite of these successes, however, a number of problems remain in several areas. In particular, the capacity of the public sector has not developed sufficiently to make implementation and enforcement of laws particularly easy. Levels of corruption are still higher than in most Western European countries, and the country is ranked only as number three in the Central and Eastern European region on Transparency International's Corruption Perception index.[2] Those governmental portfolios that were not in the forefront of the system change or of a low political priority are in a particularly disadvantageous situation and have been the main losers of the past decade and a half of change. There is certainly a case for including the Ministry of Defence in this.

This situation is further aggravated by Hungary's 'unique' military tradition. Over the past five centuries the armed forces of Hungary have achieved victory three times. Once in 1487 when King Matthias's troops occupied Vienna. A second time in 1991 when the forty member strong Hungarian medical team integrated in the British component 'won' the Gulf War. Hungary could also record victory in Kosovo as a new member of the Alliance. Beyond this, Hungary fought two world wars on the losing side and its role was reduced to that of a military springboard in the southern tier of the Warsaw Pact. In 1956 when the Hungarian people violently resisted Soviet occupation and Communist Party rule, the armed forces spontaneously dissolved and took no side in the conflict. Thereafter the Hungarian armed forces were not trusted, either by the political elite, or the Soviet leadership.

This disinterest had some surprisingly positive consequences, however. First, the defence minister was never a full member of the Politburo of the Hungarian Socialist Workers' Party (HSWP). This allowed some room for a limited professionalism ethos to develop in the armed forces. Second, the limited strategic unimportance of the country made heavy military investment unnecessary. Hungary was under much less pressure in this respect than the countries of the northern tier of the Warsaw Pact. Finally, these factors played into the hands of those who tried to carry out limited defence reform in the mid-1980s, in particular in relation to the ensuring that the military was not too great a drain on the state finances. As a consequence, defence budgets declined steadily from 1987 through to 1997.

In Hungary the system change of 1989 came about through an accord between the opposition movements/parties and the Hungarian Socialist Workers' Party (the Communist Party). Due to the nature of the change dramatic events were largely missing in the process. Hence the system change was dominated by compromise and could be characterised as a 'negotiated revolution'.[3] The 32 years of the Kádár era (1956–88), often described as the period of 'goulash communism' tried to avoid the alienation of large population groups and the 'heroization' of the tiny active opposition. The regime incorporated the overwhelming majority of the population under the slogan of 'those who are not against us are with us'. Hence it would have been contrary to the nature of Hungarian historical development had there been radical expulsions from the society on the basis of the past. Due to these factors the system change did not represent a particularly sharp divide and was not followed by 'lustration' either in the political class or in the military leadership.[4]

Democratic Control of the Armed Forces

Constitutional and Institutional Restructuring

The main constitutional and institutional changes in the political system were brought about by an extensive constitutional revision in the autumn of 1989. This has provided the framework for the development of Hungarian civil–military relations ever since. Under the modified constitution the roles of the different enforcement agencies are clearly defined. Thus, the 'fundamental duty of the armed forces (the Hungarian Army and Border Guard) is the military defence of the country'. Within the ambit of its policing activities, the border guard shall guard the borders of the country, control border traffic, and maintain order on the borders (Art. 40/A para. (1)). The 'fundamental task of the police is to maintain public safety and internal order' (Art. 40/A, para. (2)).

This institutional arrangement is similar to many other democracies. The President of the Republic 'is Chief of the armed forces' (Art. 29). The Parliament is entitled to decide on 'the declaration of a state of war and the conclusion of peace', 'the deployment of the armed forces both abroad and within the country'. It can 'establish the National Defence Council, in the case of war, or imminent danger of armed attack by a foreign power' (Art. 19, para. (3), subpara-s g), j) and h)). In peace-time it is the government that 'directs the operation of the armed forces and of the police and other security organs' (Art. 35, para. (1), subpara. h)). Partly as a consequence of political priorities in 1989, the constitutional revision emphasised the authority of the legislature over the executive in relation to the armed forces and defence policy. In practice this has had a somewhat negative impact on the effectiveness of decision-making in this area.

In particular, there are a number of constitutional constraints on the deployment of the Hungarian armed forces beyond its national borders. The Constitution stated that with 'the exception of military manoeuvres carried out according to international treaties and peacekeeping missions upon request of the United Nations, the armed forces may only cross the country's borders with the prior consent of the Parliament' (Art. 40/B, para. (1)). This rule of course deprived the country's military leadership of some flexibility from time to time, particularly since its accession to NATO in 1999. As a consequence there have been a number of attempts to try and modify this element of the Constitution to allow Hungary to become more actively and flexibly involved in the activities of the Alliance.[5] An attempted modification of the Constitution failed in December 1998 for reasons entirely unrelated to the matter of the country's forthcoming NATO membership. As a consequence, the constitution gave no room for flexibility for sending Hungarian military personnel abroad on NATO assignment. Although there was pressure to revise these rules, the government was in no hurry to do so. Indeed, it required the direct intervention of then NATO Secretary General Lord Robertson to kick start the process.[6] This led to a long awaited modification of the constitution 'in order to guarantee the carrying into effect of allied commitments' and accordingly the 'government approves the employment of Hungarian or foreign armed forces ... on the basis of the decision of the North Atlantic Council and other troop movements on the basis of the decision of the North Atlantic Treaty Organisation'.[7] This has eliminated an important obstacle to the effective interoperability of the Hungarian armed forces.

De-politicisation/De-communisation of the Military

Hungary went through a gradual transformation from communism and hence, contrary to other central and eastern European countries, no forced retirement or 'lustration' occurred in the armed forces. There would have been no particular reason for it as it was clear to all involved that the military would not resist the change. Where military resistance to reform has occurred, this has been sporadic and not of a political nature.

The constitutional and legal foundations for civilian control over the military existed early in the process of the system change. Indeed, as 'Bonapartism' has never been an issue in the modern history of the country, the core issue of ensuring civilian control over the military was not as pressing in Hungary as elsewhere in the postcommunist region. Even before 1989, the constitution stated that 'Professional members of the armed forces, the police and the civil national security services may not be members of political parties and may not engage in political activities' (Art. 40/B, para. (4)). Since then this provision has largely been adhered to, though some serving military personnel have campaigned for parliamentary office, only to resign from their posts once elected.

The Ministry of Defence (MoD) and the government were formally deprived of playing a dominant role in the command of the armed forces as part of Hungary's system for civilian control over the military. This was reflected in the separation of the MoD and the General Staff that took place on 1 December 1989. However, these changes also resulted in the development of parallel structures and a doubling of military bureaucracy. The management of military affairs was badly affected. Eventually, considerations related to effectiveness prevailed and this mistake was repaired. However, it took more than a decade to reunite the two entities and re-subordinate the General Staff to the MoD.[8] Furthermore, practitioners and insiders complain that formal ratification has been followed by painstakingly slow steps towards the merger of institutional structures. Continuing rivalries have resulted in the continuing existence of a virulent defence bureaucracy.

'Second Generation' Challenges in Civil–military Relations

Perhaps the most important issue for the quality of Hungarian civil–military relations is the nature of the interaction between civilians and military professionals in the defence sector—in particular in relation to defence planning. In practical terms, this is a far more significant issue than questions of constitutional redrafting and the 'de-communisation' of the armed forces. This situation has evolved significantly since 1990, but a number of really quite fundamental shortcomings remain.

Hungary as a new democracy could not immediately install a competent class of civilians in the MoD to support the civilian political leadership directly after the system change. The armed forces were suspicious of the few civilians who had acquired expertise in the field of defence. The fact that most Western support in training and retraining was offered to military professionals also contributed to a slow development of civilian expertise. The professional superiority of the military in defence matters, as declared by the military itself, was politically damaging. Military professionals retained significant influence on decision making inside the MoD. At the same time, the weakened civilian leadership of the ministry could not credibly represent military interests at the political (governmental and parliamentary) level. However, these difficulties did not just result from the fact that civilian expertise in defence matters was insufficient. To a significant degree, they also reflected a lack of military competence in these areas as well, particularly in relation to strategic defence planning. During the communist period, Hungary had been strategically subordinated to the Soviet High Command within the Warsaw Pact and hence had had no genuine strategic culture of its own for at least four decades. After the system change, these deficiencies meant that the Hungarian military itself was significantly lacking in expertise in these areas, and was ill-equipped to shape the country's new defence policy. As a consequence, for much of the 1990s, Hungarian civil–military relations were characterised by two largely incompetent groups facing each other: the new civilians in the defence sector and the old military 'professionals'.

In the 15 years that have passed since the system change, these challenges of coexistence and cooperation have resulted in a number of tensions. The civilians have often looked down on the military professionals and tried to reduce their autonomy, even in those areas where they had clear competence—such as military-technical matters. The military in turn have tried to present as many issues as possible as falling within their own sphere of exclusive competence.

This 'cats and dogs game' has been going on for some time with the strength of one group *vis-à-vis* the other varying over time. At certain times the increasing influence of the military has led to the 're-militarisation' of the defence establishment whereas on other occasions civilians influence was maximised. The period of the Socialist-Liberal Horn, Medgyessy and Gyurcsány governments (1994–8 and 2002–4 and 2004–, respectively), for example, were characterised by a process of 're-militarisation' whereas the two conservative cabinets (1990–4, 1998–2002) proved more determined to civilianise the defence ministry. The Socialist-Liberal position has been understandable, although not acceptable, and results from the traditionally good relationship that the Socialist Party maintained with the armed forces during the 'communist' period. In the main therefore, the process of developing the professional civilian component of Hungary's defence sector has been set back each time a Socialist-Liberal government has been elected. Despite this, however, there is a chance that increasing competence on both sides of the civil–military divide in the Hungarian defence sector may result in a more balanced relationship between the leadership of the MoD and the General Staff in the long run. Partly as a consequence of these difficulties, Hungary has continued to lag behind international expectations in relation to its defence reforms. Perhaps most importantly, it has been slow to develop a modern military establishment, conscious of its tasks and responsibilities and with a clear vision for the future.

Human factors have also played an important role. For the first 12 years of the Republic of Hungary's existence, the post of defence minister tended to be offered as compensation or part of a political deal. As a consequence, the persons selected for the post often had no competence or even particular interest in this field. Moreover their position was generally weak in the political establishment more broadly. For example, the first minister, Lajos Für (1990–4) was appointed to the post in compensation for not being elected president of the republic. György Keleti (1994–8) became defence minister because he had been the press spokesman of the MoD and had been directly elected in the first round of the elections with a convincing majority. Although he had no proper military experience (he used to work in press matters and at the party committee of the MoD) he had fair knowledge of the ministry. His own interpretation of his role was to contribute to the Hungarian state's budget balance and not to present excessive financial demands to the Finance Ministry. János Szabó (1998–2002) became the Minister of Defence at the last minute. He belonged to the junior coalition partner in the new government—the Smallholders' Party—which demonstrated mass incompetence across all the portfolios for which it was responsible. In addition, on many occasions, the Prime Minister systematically violated the decision-making autonomy of the Minister of Defence enshrined in the Law on National Defence.

In the wake of the 2002 parliamentary elections, this situation has at last changed. Even before the elections took place, it was clear that irrespective of which major party formed the government, the Minister of Defence would be a politician with a genuine interest in defence matters for the first time. So it happened. The new Socialist Minister of Defence, Mr Ferenc Juhász had been a member of the parliamentary defence committee for eight years, and its vice-president for four. In spite of this, his lack of bureaucratic experience has handicapped his performance and in practical terms his achievements have been small. The defence minister faced three major challenges during the first two years of his office term: first, defence reform; second, tackling corruption in the defence sector; and finally fulfilling his personal aspirations. In each case the results have been disappointing: defence reform has eaten up much time without bringing the long awaited breakthrough in the performance of the Hungarian armed forces; rumours persist that corruption has not continued in the MoD and the Defence Minister himself has been implicated; and Juhász's own ambitions to become leader of the Socialist Party have proved to be unrealistic. As a consequence, it seems that the chaos of the defence sector will continue in the years to come.

The role of the Hungarian parliament in defence matters has been very similar to its equivalents in the better established democracies. The difference is not in the formal rules governing parliamentary oversight, but rather in the practical functioning of these bodies. For example, despite the fact several MPs have held their positions since the early days of Hungarian democracy and have been members of the defence committee for more than one legislative period, their competence in defence matters have not reached desirable levels. This has been due partly to the relatively little attention paid to defence matters by both the government and parliamentarians. This in turn has reflected both the largely threat-free geostrategic environment that Hungary finds itself in, as well as Hungarian society's traditional lack of interest in defence issues.[9] Nevertheless, Hungary's increasing participation in international military operations—over which parliament had and to a lesser extent still has real influence—has recently forced defence matters higher up the parliamentary agenda.

Professionalisation and Military Reform

When Hungary started its transition in the late 1980s it was clear there were many other items on its political and economic agenda that were far more important than the transformation of the defence sector. As far as the new political elite were concerned, the key issue relating to the military was that it would not intervene in the political process. Once it became clear that this was uncontestably the case their interest in further defence reform issues declined significantly and priority was instead given to the issue of institution building. The Antall government—Hungary's first after the system change— performed very well in this respect and made good progress in establishing civilian control over the military and clarifying constitutional arrangements for civil–military relations. In addition, the Antall government was also clearly—though understandably at this point primarily symbolically—committed to the idea of the NATO integration and made no secret of the fact to its Western partners. Understandably, this commitment remained primarily symbolic for most of the early 1990s and NATO itself gave little indication of its willingness to accept Hungary into the Alliance. As a consequence, it was enough for Hungary to demonstrate a strong desire to join NATO at the levels of declaratory policy and diplomacy alone and it did not translate into practical defence reform or modernisation efforts. In contrast, the launch of the Partnership for Peace (PfP) programme during the last months of the new government's office in 1994 offered the real prospect of potential NATO accession at some future point. Significantly, PfP also made clear that the modernisation of applicants' defence sectors would be an indispensable precondition of NATO integration, if not necessarily of membership.

When the Socialist-Liberal Horn government came into power in 1994, it was preoccupied with more pressing matters than defence reform. These included the consolidation of the economic situation, including the introduction of the so-called Bokros austerity package in March 1995, and resolving foreign policy tensions with Romania and Slovakia. In relation to the defence sector, the Horn cabinet apparently started out from the mistaken assumption that NATO accession would not occur any time soon. Hence, they believed that the best contribution the defence sector could make to the country's reform process would be to ensure that it did not undermine the shaky balance of the national economy. In terms of policy they followed the pattern of their predecessors, further shrinking the defence budget and introducing different military reform plans so swiftly that they had no practical chance of putting them into practice. The few important steps were taken were done so in order to meet the expectations of the world at large, and particularly those of NATO and its member-states. These included the establishment of a coordination mechanism to prepare for NATO accession. After NATO actually issued its invitation to Hungary in the summer of 1997, the Horn government demonstrated little more determination in relation to reforming the defence sector. This was because it believed the accession process would be motivated primarily on the basis of a political decisions rather than Hungary's military-technical performance.[10] In addition, by this time new elections were looming which served to delay the introduction of any major new policy initiatives in this area.

A new government—under Prime Minister Viktor Orbán—was formed in the summer of 1998. However, despite the fact that Hungary was due to accede to NATO only a few months later, defence reform initially remained a low priority. This was primarily because of the continuing lukewarm attitude of the majority of the population towards defence matters and the consequent political prioritisation of other policy areas. Nevertheless, the Orbán government did initiate two important positive developments in relation to Hungary's defence reforms. First, it played a constructive and cooperative role with other members of NATO during the Kosovo operation in Spring 1999. Second, after the conflict it introduced a long-term and fairly comprehensive defence reform plan, which despite certain shortcomings was practical to implement.[11]

For two reasons it had become clear that Hungary's defence reform could not be postponed further. First, considerable pressure had built up over the need to transform a defence structure which absorbed major resources, without contributing greatly to the defence capability and the international reputation of the country. The disruption caused by the Hungarian government's decision to send one battalion to the NATO-led Kosovo Force (KFOR) in July 1999, for example, highlighted the financial fragility of the Hungarian defence budget. The second pressure for reform was international, and stemmed from the need for the government to demonstrate it could shoulder the burden of NATO membership. Specific Hungarian deficiencies were highlighted during the Kosovo crisis, when, after Serbian planes had violated Hungarian airspace, the air force had to rely on NATO allies to patrol its airspace due to the absence of NATO standard Identification Friend or Foe system on Hungarian Mig 29s.

In the summer of 2000 the parliament approved a plan for the long-term transformation of the Hungarian Defence Forces (HDF). The process was divided into three phases between 2000 and 2010. During the first phase the emphasis of the reform was on the transition to the new structure, relocation of troops, the establishment of adequate proportions of personnel strength, the creation of the basis for reducing operational costs and costs of maintenance, improvement of living and working conditions, and the establishment of a minimum level of NATO interoperability. The second phase from 2003 to 2006, would pursue programmes to improve the quality of life, combat capability and training of the HDF. Until the end of this phase the armed forces would essentially operate on the basis of existing—though in some cases upgraded — equipment. Phase three from 2006 to 2010 aimed at the modernisation of equipment in accordance with capability requirements, and the demands of increased NATO compatibility and interoperability. Unfortunately however, this fairly comprehensive plan had no chance to be implemented as the Orbán government lost the elections in the spring of 2002.

From the beginning of the new government's period in office, it was obvious that the sharply divided establishment would take over as little of the legacy of the conservative cabinet as possible. There were two apparent shortcomings of the defence reform plan. Namely, it intended to maintain a mass army, even though much smaller than before. This was closely associated with the insistence of the Orbán cabinet upon maintaining conscription for a longer period of time. For his own ideological reasons, Orbán himself was determined to maintain conscription. For many, Orbán's reasoning on this issue was difficult to fathom, particularly since the service time of conscripts had been reduced to only six months at the beginning of 2002, which made any military utility they may have had highly questionable. Furthermore, conscript soldiers were constitutionally forbidden from participating in any international missions. This meant that in practice there were two separate armies inside the HDF. In the framework of selective development a few units were prepared for international operations whereas the others were sustained on a low level of preparedness and equipment. This was understandable temporarily and meant that Hungarian contributions drawn from elite units performed fairly well in international operations such as Bosnia and Kosovo. In practical terms this policy meant that Hungary combined an unreformed mass army with some pockets of a modern armed force.

One of the major shortcomings of the Orbán government's period in office was that they did not foresee that NATO accession would greatly increase the foreign policy significance of the defence sector and raise the importance of the MoD as a conduit for cooperation with the Alliance. The new significance of the MoD required a leadership that in particular would be able to communicate the (few) achievements of the Hungarian defence sector to NATO and its members and also explain and generate understanding of its shortcomings. In practice, however, the MoD leadership was not up to this important task, a situation that resulted in poor communication with NATO at the highest level and complemented the generally weak performance of the Hungarian defence sector within the Alliance.

When the Socialist-Liberal coalition government was formed under Prime Minister Péter Medgyessy in May 2002 there were hopes that the professional competence of the new Minister of Defence would make a difference in this area. Indeed, one of his first decisions on coming into office was to order a review of the defence capabilities of the Hungarian defence sector. Perhaps most significantly, he announced a plan to abolish conscription in 2005 and further reduce the number of military facilities.[12] This is a major difference from the policy of the Orbán government as it means that conscription will be abolished during the term of office of the current administration. While this may still be too little too late, it *will* mean that the HDF's 'two armies' will gradually shrink to one and that professionalism in the entire armed forces will prevail. The current reform process does still remain threatened by potential political compromises and 'technical failures' however. In relation to the latter, for example, the successful establishment of a clear career structure for professional officers is of great importance. If this does not occur, then it is likely that the younger, more active, and competitive military professionals— who are the HDFs brightest hope for the future—will quickly leave the armed forces. In their place, the military risks being left with 'an armada of lieutenant colonels'.[13] Nevertheless, despite all the potential shortcomings associated with abolishing conscription and establishing professional armed forces, this policy initiative may well prove to be the single most important achievement of the Socialist-Liberal government's defence reform programme. Given the gross unpopularity of compulsory military service in Hungary it is also a step that will most likely prove to be irreversible.

As well as implementing its plans for abolishing conscription, the Medgyessy-led coalition has also introduced its own three phase defence reform plan. The first phase, which will run until the end of 2006 will concentrate on personnel and training reform in the HDF. This will involve closing several units and facilities (including the Pécs artillery brigade, the training facilities at Szabadszállás and the Savaria training centre at Szombathely) and relocating others (such as the air defence missile regiment and the combat helicopter regiment).[14] There will also be a reduction in the number of officers, which will have a positive impact on the officer to NCO ration in the HDF. In addition, the reform plan will concentrate on properly preparing and equipping those units which will be offered to NATO assignments, and redundant armament and materiel will be withdrawn from service. During the second phase, between 2006 and 2010, the modernisation process will continue. Fighting units will continue to be re-equipped with new vehicles, armaments and materiel while the number of military facilities will be reduced by a further 20%. In phase three, between 2010 and 2013, the reform envisages the completion of the qualitative development of the HDF—including the living and working conditions of military personnel—to full NATO standards.

Debates continue to surround the elaboration of the defence reform. Opposition parties, which still attribute more importance to the individual defence of national territory than the Socialist-Liberal government, insisted upon maintaining organisations of mobilisation and territorial defence. This is an option that has been rejected by the government, partly due to the clear absence of external threat, and partly due to Hungary's membership of NATO which it argues makes the individual self-defence of national territory unnecessary. The two opposition parties are also of the view that the amount Hungary is prepared to commit to peace operations should be determined as a percentage of the defence budget. This would in practice limit the amount that could be spend on peacekeeping missions for the HDF, while protecting resources for their 'core' mission of individual and collective self-defence. This proposal was rejected by the government on the basis that it violated its legally enshrined right to approve the HDF's participation in NATO peace operations.[15] Perhaps more significantly, however, had Hungary's ability to contribute to peace operations been constrained it would have deprived the country of its most valuable contribution to international security and its participation in the Atlantic Alliance. In contrast to the military's practically non-existent defence of national territory role for example, there are currently 964 Hungarian personnel on international assignment. The four largest missions are the transportation battalion and headquarters component in Iraq (295), KFOR (266), SFOR (142) and UNFICYP in Cyprus (122).[16]

Despite these reforms however, there appears to be little understanding in the MoD leadership that the changes associated with professionalising the HDF might require a different philosophy than before. The previous practice of differentiating between units and investing in some of them at the expense of others cannot be sustained for example. Furthermore, the communication problems that characterised the MoD under the Orbán government have continued, albeit on a somewhat reduced scale. This burdens communication with NATO and has increased the external pressure that has been mounting on the Hungarian defence sector over the past two years. Similarly, while the government has committed itself to increasing military expenditure from 1.71% of GDP in 2004 to a 1.81% in 2006 (to be sustained for the following 10 years)[17], in practice this may not be enough to significantly ease the defence sector's current constrained financial situation. Moreover, as Sebestyén Gorka stated in 2004, 'the root problem is not funding but waste and corruption. The budget is big enough to create a decent army if it were managed properly.'[18]

The Military and Society

The prestige of the armed forces changed significantly since the system change in 1989. This is partly a consequence of the regained independence of the country, which has made the armed forces more of a national institution than before. Hungarian military–society relations remain coloured by the fact that Hungary has no rich military tradition and Hungarians have no tradition of engaging either widely or deeply in defence matters. In practice this means that support for the armed forces in Hungarian society tends to be rather volatile. Thus, when the armed forces' activities are publicly visible, their popularity in society tends to rise. This was the case when they played a key role in fighting the devastating flooding which afflicted the country in 1999 and 2000, as well as before and during NATO's Kosovo operation. In the absence of such events, they generally receive little societal attention or support.[19] The legitimacy of the military in Hungarian society today therefore tends to depend on it being able to constructively fulfil tasks that are perceived to be worthwhile.

It is unlikely that the introduction of a fully professional army by the end of 2005 will 'detach' the Hungarian armed forces from the society. Indeed, in many respects it is likely that Hungarian society would be prouder of a smaller, properly functioning, professional force, than the malfunctioning conscript army of the past 15 years. It is not therefore necessary to compensate for the abolition of conscription by creating territorially composed armed units analogous to the US National Guard. While military service persists, it remains deeply unpopular. Draft dodging continues to be a severe problem and is likely to remain so for the remaining period until the final batch of conscripts are called upon to serve. Conscription in Hungary in its present form is both unworkable and politically unpopular. As a consequence it is likely that its abolition will be welcomed not only by those who are eligible for the draft, but also by the political establishment and the military leadership.

In the absence of radical and unexpected change Hungary can expect to exist in a relatively benign international environment for some time to come. In this regional context, military force is likely to be used primarily for conflict management and peacekeeping operations, and for coercion only exceptionally (and then generally only by the great powers). As the integrated space of Europe expands there is reason to hope that a genuine democratic peace will prevail. This would mean that the challenges to the Hungarian military will appear hundreds, or more likely thousands of kilometres away from the borders of the country rather than on their doorstep as in the past. In such a situation it will continue to be a challenging task to maintain societal support for the defence sector and particularly to guarantee adequate financing.

The current Medgyessy government has promised to adequately fund the armed forces and particularly to reverse the consistently declining funding available for technical modernisation. However, in a benign regional security environment, defence remains a vulnerable policy sector, especially in comparison to those areas which have a more direct bearing upon the economic competitiveness of the country or that influence voting patterns more directly. The armed forces continuing weak legitimacy in society therefore mean that it is likely that they will be the first institution whose funding will be cut in the event of scarcity of state resources. This was already a problem for the defence sector in 2004, when the financial pressures required major cuts in the state budget. The defence ministry was an easy target and lost approximately 12 billion Hungarian forints (US$ 60 million). In real terms this meant a reduction of approximately 3% in the defence budget for 2004. Scarcity of resources in combination with waste and corruption in the system is a damaging mix that may undermine the performance of the armed forces in the future.

International and Transnational Influences

It is extremely difficult to carry out major, sustained military reform when there is no external threat that would support the allocation of resources to this sector, and when available human resources are limited. Very few countries of the world can escape this dilemma, which applies across the entire spectrum of military activities. Selective development in certain key areas may help but it would require broad consensus in the Hungarian political elite to agree upon those key areas where such development should concentrate.[20] Bearing in mind that since the system change each democratically elected government of Hungary has served its office term and none of them have been re-elected, it is also impossible to carry out military reform without broad political consensus. In light of the mutual alienation of the main political forces due to the populist rhetoric of former Prime Minister Orbán and his entourage Hungary will most probably continue to be a difficult case for the rest of NATO, at least as far as its defence sector is concerned.

Hungary's defence reform performance since it has joined NATO has primarily generated disappointment amongst its allies. The United States in particular has been vocal in its criticism. However, to date, Hungary has successfully managed to compensate for its limited reform progress within the Alliance through four main strategies. First, Hungary has been an extremely *loyal* member of the Alliance and has not interfered with the organisation's decision-making process in any measurable way. This position was most clearly visible during the Kosovo operation when decisions were carried out as smoothly with 19 members as they would have been with 16. Hungary also contributed military units to the two major NATO-led peacekeeping operations in the Balkans, SFOR and KFOR. Hungary was also one of the eight states which rallied to the support of the UK and US in 2003 over the issue of military intervention against Iraq.[21] Second, Hungary has played an important role within the Alliance simply because of its *location*. Indeed, its position as a neighbour of three of the successor states of the former Yugoslavia including Serbia meant that it was in an especially important position during the Kosovo crisis, providing NATO full access its airspace, airfields and other military facilities whenever necessary. Similarly during the Anglo-American operation in Iraq, Hungary opened its airspace for transit and permitted the training of Iraqi opposition personnel at the Taszár airbase.[22] More recently, it expressed its readiness to allow up to 28,000 Iraqi policemen to be trained at the same base. Third, partly because of its strategic location, and partly because of its familiarity with its more strategically important neighbours, Hungary has helped to contribute to the common knowledge of the Alliance. It has also actively cooperated in the field of *intelligence*. Finally, Hungary has managed to compensate for its weak performance in defence reform through the *promises* it has made. It has been extremely skilful in making promises and seldom delivering on them later.[23]

Despite these strategies, NATO pressure on Hungary over the defence reform issue has been constant and sometimes public. As a consequence, the government has gradually recognised that there are a number of defence reform issues that it cannot escape from. These include the reduction in the size of the armed forces, the planned reallocation of resources from personnel to investment, equipment modernisation for the HDF, and increasing the proportion of its GDP that is spent on defence. Indeed, on current plans, the HDF will be downsized to less than 30,000 people, will invest more than at any time since 1987 and the defence budget will grow to 1.81% of GDP by 2006.[24] Despite the fact that Hungary originally promised to achieve these goals even before it acceded to NATO—late delivery is better than no delivery at all. It is also clear that this would not have been possible at all without steady external pressure on the government to do so.

The changes will result in a situation where the Hungarian armed forces will continue to be able to contribute effectively to international operations. The reorientation of the armed forces has taken place under the assumption that the geostrategic position of the country will not change significantly and that Hungary will not have to take care of its own defence individually. If this set of assumptions proves to be unfounded Hungary will have to reconsider its defence fundamentally. If on the contrary the projections will be realistic then the conclusion can be drawn that a good part of the way has already been taken on the road to defence modernisation.

Notes

[1] The two countries that represent the higher intra-EU export share are the Netherlands and Portugal. Several EU member-states have higher *volume* of intra-EU trade than Hungary, of course. It is remarkable, however, that after the 1 May 2004 enlargement more than 82% of Hungarian exports will be intra-EU.

Transformation of the Hungarian Military 31

[2] Hungary used to be second on the list but slipped to third in 2002. Globally, Hungary ranked 31 on the list in 2001, behind Trinidad and Tobago and with Tunisia and only Estonia preceding it from among the transition countries of East-central Europe. In 2002 Hungary slipped to rank 33 together with Malaysia and Trinidad and Tobago, preceded by Slovenia (rank 27) and Estonia (rank 29). In 2003 this trend continued. Hungary ranked number 40 whereas in 2004 number 42. Interestingly its ranking in East–Central Europe did not change. See Transparency International Corruption Perceptions Index 2001, 2002, 2003 and 2004.

[3] R. L. Tőkés, *Hungary's Negotiated Revolution* (Cambridge: Cambridge University Press, 1996).

[4] There has been a certain 'soft lustration' of the political elite in the sense that those who had held certain political functions or had had access to information of the internal counter-intelligence were screened. If they were found to have held certain political functions they had two choices. Either they had to give up their function or their position would have to be made public. This did not prevent the electorate from voting people who had been affected by this rule into the highest offices.

[5] Lányi Zsolt felszólalása az Országgyűlés 1998. december 2-i ülésén /The contribution of Zsolt Lányi at the 2 December 1998 session of the National Assembly/. Available at Http:// www.mkogy.hu/internet/plsql/ogy_naplo.naplo.'fadat?p_ckl = 36p_uln = 38p_fe (accessed on 6 September 2004).

[6] In autumn 2003 Lord Robertson publicly named Hungary as a country whose constitutional rules did not allow it to deploy its armed forces on NATO missions. The somewhat unusually specific reference to one NATO member-state prompted a reconsideration of the issue.

[7] 2003. evi CIX. törvény a Magyar Köztársaság Alkotmányáról szóló 1949. évi XX. Törvény módosításáról /Act CIX of 2003 on the modification of Act XX of 1949 on the Constitution of the Republic of Hungary/. Available at http://www.complex.hu/kzldat/t0300109.htm/t0300109.htm (accessed on 6 September 2004).

[8] For details of this fascinating story see, 'Civil–military Relations in Hungary: No Big Deal', in Andrew Cottey, Timothy Edmunds & Anthony Forster (eds), *Democratic Control of the Military in Post Communist Europe: Guarding the Guards* (Houndmills: Palgrave, 2002), pp. 64–87.

[9] The potential threats emanating from the former Soviet Union, in contrast with many other Central and Eastern European countries, have never had any measurable impact upon the threat perception of the Hungarian public. For details see Ferenc Molnár, 'A közvélemény alakulása a biztonságról és a haderők szerepéről a Cseh Köztársaságban, Lengyelországban és Magyarországon', *Új Honvédségi Szemle*, 54: 8 (August 2000), pp. 4–23.

[10] In this respect the Horn government was by and large right. The first post-Cold War enlargement of NATO did not pay particular attention to military preparedness. When at a late stage it did, it simply required that the applicant states increased the share of the state budget that they spent on the military. This fiscal approach did not require any actual improvement of military capabilities, however.

[11] For more details concerning the reform see 'Hungary', in H. J. Giessmann & G. E. Gustenau (eds), *Security Handbook 2001* (Baden-Baden: Nomos Verlag, 2001), pp. 249–82, 'Hungary: Peace and Quiet of an Increasingly Illiberal Democracy', in D. N. Nelson & U. Markus (eds), *Brassey's Central and Eastern European Security Yearbook* (Washington D.C.: Brassey's, 2002), pp. 85–119, 'Building Professional Competence in Hungary's Defence: Slow Motion', in A. Forster, T. Edmunds & A. Cottey (eds), *The Challenge of Military Reform in Postcommunist Europe* (Houndmills: Palgrave, 2002), pp. 63–78.

[12] Béla Szilágyi, 'Csak 2005-ben szerel le az utolsó sorkatona' /The last conscript soldier will be discharged in 2005 only/, *Magyar Hírlap*, 13 September 2003.

[13] See Gábor Miklós, Alezredesi ámádia /An armada of lieutenant colonels/, *Népszabadság*, 6 June 2003.

[14] See A kormány jóváhagyta a haderőreform terveit /The government approved the military reform plans/, 25 September 2003. Available at http://www.honvedelem.hu/cikkphp?cikk = 14241 (accessed on 6 September 2004); *Népszabadság*, 31 July 2004, p. 7.

[15] Iváncsik Imre honvédelmi államtitkár bevezetö elöadása a Magyar Honvédség hosszú távú fejlesztésének irányairól valamint a Magyar Honvédség részletes bontású létszámáról szóló országgyűlési határozati javaslatok együttes vitájában /Introductory statement of Imre Iváncsik, political state secretary of the MoD, in the joint debate on decision of Parliament on the long term directions of the development of the Hungarian Defence Forces and detailed headcount of the personnel of the Hungarian Defence Forces/ 2 March 2004. Available at http://www.mkogy.hu/internet/plsql/ogz_naplo_fadat_aktus?p_ckl=37&p_uln=12 (accessed on 6 September 2004).

[16] Magyar katonák szerepvállalása a világ békéjének és biztonságának megteremtésében /The role of Hungarian soldiers in creating peace and security in the world/ Last updated on 10 June 2004. Available at http://www.honvedelem.hu/missziok_ index.php (accessed on 20 June 2004).

[17] A kormány jóváhagyta a haderöreform terveit /The government approved the military reform plans/, 25 September 2003. Available at http://www.honvedelem.hu/cikkphp?cikk=14241 (accessed on 20 March 2004).

[18] Christopher Condon, 'Top brass needs to raise its game: Military spending is likely to rise from 1.5 per cent of GDP but critics say money will be wasted', *Financial Times*, Special Report: Hungary, 1 June 2004, p. 4.

[19] See the public opinion poll data presented in Pál Dunay 'The Armed Forces in Hungarian Society: Finding a Role?', in Anthony Forster, Timothy Edmunds & Andrew Cottey (eds), *Soldiers and Societies in Postcommunist Europe: Legitimacy and Change* (Houndmills: Palgrave-Macmillan, 2003), pp. 84–9.

[20] The list Hungary presented at the Prague NATO summit in late November 2002 included the improvement of the mobility of the Hungarian Defence Forces, defence against biological and chemical weapons, logistical support to the deployment of the forces of allied powers as well as to guarantee the air refuelling capability of the Gripen aircraft Hungary has purchased. It was not made clear whether Hungary would purchase An-70 transport aircraft or transport capacity will be established in cooperation with other countries. See '30 milliárd forintos magyar felajánlás a NATO-nak' /30 bn forints commitment to NATO/. Available at http://www.korridor.hu/cikk.php?cikk=100000044877 (accessed on 17 March 2004).

[21] See P. Webster, 'Eight leaders rally 'new' Europe to America's side', *The Times*, 30 January 2003 and A. Applebaum, 'Here Comes the New Europe', *The Washington Post*, 29 January 2003.

[22] Az iraki partoknál a Taszáron kiképzettek egy csoportja /A group of those trained at Taszar are at the shores of Iraq/ 27 March 2003. Available at http://www.honvedelem.hu/Popup_index.php?type=nyomtat=id (accessed on 30 March 2004).

[23] As defence minister Juhász said in an interview the day before the coming into office of the new government 'following NATO accession the country fulfilled 76 per cent of its pledges, 50 per cent a year later and only 26 per cent in 2001'. See Mihály Bak, 'Nem minden vezetö marad a helyén a honvédelmi tárcánál—Négyszemközt Juhász Ferenccel' /Not every leader keeps its post at the defence portfolio—Eye to eye with Ferenc Juhász/, *Magyar Hírlap*, 26 May 2002.

[24] L. 'Védelmi Felülvizsgálat : Uton a XXI. század hadserege felé'. Available at http://www.honvedelem.hu/cikk.php?cikk=13776 (accessed on 18 March 2004).

The Transformation of Postcommunist Civil–military Relations in Poland[1]

PAUL LATAWSKI
Department of Defence and International Affairs, Royal Military Academy, Sandhurst

For postcommunist Poland, the problems associated with the transformation of civil–military relations have been significant, but have in no way diverted the fundamentally successful trajectory of the country's transition toward a consolidated democracy. The Polish case shares many of the characteristics of other former communist states in terms of the immediate legacy of communist civil military relations. Nevertheless, the story of postcommunist civil–military relations has also been shaped by current features found in its democratic transition as well as deeper legacies found in Poland's diverse political and military historical experience. The aim of this paper is to examine the transformation of civil–military relations in Poland since 1989. In doing so, it will consider civil–military relations in its broadest sense, so as to embrace the issues of the establishment of democratic control of the armed forces, professionalisation of the military and the relationship of military and society. In examining these three strands of civil–military relations, the paper will assess the major factors influencing the changes in Polish civil–military relations.

Democratic Control of the Armed Forces

At the core of the transformation of Polish civil–military relations is democratic control of the armed forces. For Poland's transition to a consolidated democracy, the attainment of democratic control is a *sine qua non* of its political life. It has, however, been one of the areas that has generated doubt and discussion about the prospects for the establishment of democratic-civilian control over the armed forces. Despite the uncertainty, initial impressions may be misleading. In broad terms, it is clear that democratic control certainly exists both in principle and in practice, even if why it works can appear mysterious to outside observers. This uneasy perspective is the product of two things: the weakness of political society in the management of defence and the scale and pace of constitutional-legal-institutional reform to be undertaken.

The introduction of changes to the constitutional and legal frameworks and the shaping of institutional design was a slow and deliberate process. Changes to the 'rules' and institutions, however, are vital to a more effective approach of political society to the management of democratic control of the armed forces. Between 1989 and 1991, the changes made to the constitutional-legal order and institutions in Poland reflected the immediate need to make provisional changes in line with the shift away from a communist to a democratic state. The most important priority was to make initial steps in establishing control of Poland's defence establishment. The amendment of the 1967 statute, Law on Duty to Defend the Republic of Poland, published in the Journal of Law (*Dziennik Ustaw*) on 11 December 1991 outlined a system for controlling the armed forces. The modified law ostensibly gave the President the dominant influence over the armed forces. Article 25 of the Law made the President 'Commander-in-Chief of the Armed Forces' and gave him the authority to determine their 'main directions of development' but 'on the recommendation of the Minister of Defence'. The tone if not the intent of the 1991 law clearly gives a *primus inter pares* position to the President in controlling and guiding the defence sphere but ambiguities remained regarding the roles of President and Minister of Defence.[2] For example, the 1991 statute stated that the Defence Minister 'Commands the Armed Forces of the Republic of Poland' that seemingly duplicated the President's role as commander-in-chief.[3] The elimination of the lack of clarity in roles, however, ultimately depended on changes to Poland's constitution.

By December 1992, the adoption of a provisional 'small' constitution did not resolve the ambiguity. The 'small' constitution was a heavily amended version of the inherited communist one and was only ever conceived to be a stopgap. Article 34 of the 'Small' Constitution stated, as in the case of the 1991 statute law, that the 'President shall exercise general supervision with respect to the external and internal security of the state'. However, Article 52 of the 'Small' Constitution also indicated that the Council of Ministers 'shall ensure the external and internal security of the state'. Moreover, Article 35 (1) made the President 'Supreme Commander' of the Polish armed forces, but this role was not adequately defined either in the Small Constitution or in statute law. The

President also had to be consulted by the government in the appointment of the Defence Minister.[4] The interim constitutional changes (and the preceding changes to statute law) created a tension between the President and Defence Minister in exercising, both in theory and practice, democratic control of the armed forces.

The new Polish Constitution adopted in 1997 did much to add clarity to the norms and machinery of state charged with democratic control and oversight of the armed forces. Article 26 of the Constitution provided a normative benchmark for civil–military relations: 'The Armed Forces shall observe neutrality regarding political matters and shall be subject to civil and democratic control'.[5] Significantly this article uses the term 'neutrality' regarding politics rather than 'apolitical'. Neutrality does not suggest direct involvement in politics nor does it preclude the political rights of service personnel as individual citizens. In contrast 'apolitical' could be misconstrued as meaning institutional separateness or being outside political control.

The 1997 Constitution shifted the executive balance toward the Defence Minister (Council of Ministers) and away from the President. Article 134 (2) of the constitution stated that the Defence Minister 'shall exercise command over the Armed Forces through the Minister of National Defence'. Functions that the earlier 'Small' Constitution assigned to the President now became the responsibility of the Council of Ministers in Article 146:

- exercise general control in the field of national defence and annually specify the number of citizens who are required to perform active military service
- ensure the internal security of the State and public order;
- ensure the external security of the State.[6]

The President's constitutional role regarding defence was seemingly little diminished from the preceding 'Small' Constitution as Article 134 (1) stated that: 'The President of the Republic shall be the Supreme Commander of the Armed Forces of the Republic of Poland'. However, the elaboration of this role was left to Parliament and the Council of Ministers in the 1997 Constitution in Article 134 (6). Throughout Article 134 of the 1997 Constitution, the President's roles vis-à-vis the armed forces, including appointing a commander-in-chief in wartime, making senior military appointments or conferring ranks requires either the concurrence of the minister of defence or procedures to be 'specified by statute'.[7] The President is, however, more than a figurehead on defence matters. In addition to adding his voice to public debate on defence issues, under the 1997 Constitution the President can convene a 'Cabinet Council' where he chairs a meeting of the Council of Ministers or deliver a message either to a joint or individual sitting of both parts of the Parliament.[8] The key working relationship on defence matters remains between President and Defence Minister. This relationship depends on the enactment of the statutes defining it according to the 1997 Constitution. There can be little doubt, however, that the dominant executive figure is the Defence Minister.

Institutional Evolution

The institutional development of the national security system has reflected the tensions between the competing models of government and presidential executive control. During the period of the Small Constitution, the institutional development tended to reflect stronger presidential prerogatives. The most important institution, apart from the Ministry of Defence, was the National Defence Committee (*Komitet Obrony Kraju* (KOK)). Inherited from the communist period, the 1991 statute, Law on Duty to Defend the Republic of Poland, gave this body and the President wide powers for shaping Polish defence and security policy.[9] Moreover, it effectively gave KOK law-making powers regarding the armed forces.[10] The President also acquired in this statute a National Security Office (*Biuro Bezpieczeństwa Narodowego* (BBN)) to act as a secretariat for the KOK with the head of the BBN serving as the secretary to the KOK.[11] The membership of KOK brought together the President, Prime Minister, key ministers and figures from the Parliament. The role of the KOK and BBN *vis-à-vis* the Council of Ministers and the Ministry of Defence was bound to muddle lines of responsibility.[12] By 1994, the Council of Ministers effectively developed a parallel institution to the KOK. The creation of the Committee of Defence Affairs (*Komitet Spraw Obronnych Rady Ministrów* (KSORM)) provided the Council of Ministers with a forum to develop defence policy at an inter-ministerial level.[13] These institutional arrangements only mirrored the lack of a clear constitutional steer on whether the President or the Council of Ministers had primacy on defence matters.

The adoption of the 1997 Constitution promised to resolve the institutional duplication. On the face of it, however, the new constitution only seemed to perpetuate the proliferation of institutions. It created under Article 135 yet another defence related institution—the National Security Council (*Rada Bezpieczeństwa Narodowego* (RBN)). The RBN's function was to be: 'The advisory organ to the President of the Republic regarding internal and external security of the State shall be the National Security Council.'[14] It was, however, the only one of the plethora of security institutions that had constitutional standing. All the others were creatures of statute law. The membership of the RBN is virtually identical to the KOK with the exception of the replacement of the Chief of the General Staff by the President of the National Bank.[15] The provisions of the 1997 Constitution with its decisive shift in favour of the Council Ministers having the dominant executive role in controlling the military meant that one or possibly two of the existing institutions were now destined to be eliminated. The most important of these was KOK. The statute law that underpinned KOK, particularly from the point of view of presidential influence conflicted with the new constitution. The question that quickly emerged after the new constitution came into force was not whether KOK should disappear but how its powers would be distributed.[16] The institutional shake out that followed the 1997 Constitution streamlined the management of defence and saw KOK and KSORM eventually disappear.[17]

Since the early 1990s, successive Defence Ministers have introduced a series of reforms designed to streamline its military components and 'civilianise' major elements of the ministry of defence. In 1993, the then Polish Minister of Defence, Janusz Onyszkiewicz attempted to introduce a new set of regulations and to reorganise the Polish Ministry of Defence, but this effort fell foul of political difficulties and the obstacles posed by the duality of executive control under the Small Constitution.[18] Subsequent attempts to reorganise the Defence Ministry proved more successful and by January 1996 the Ministry underwent a thorough reorganisation. The authority of the Defence Minister was enhanced and carefully outlined in a statute dated 14 December 1995. The new statute established an unambiguous chain-of-command to subordinate the most senior military officer. Article 7.1 states that 'the chief of the General Staff of the Polish Army is directly subordinated to the minister of national defence' and it goes on to say in Article 9.2 that 'the decisions of the minister of national defence have the force of a military order'.[19] All of these changes strengthened civilian control of the armed forces.

Oversight of Defence

The establishment of oversight of Polish defence proved to be a more straightforward development. The most important institution for this function is the Parliament (Sejm). Its control over the budget of the armed forces, power to legislate and ability to issue a vote of no confidence in the government, are undoubtedly powerful instruments in a democratic civil–military relationship, but its ongoing oversight function is perhaps the most important element in the Parliament's relationship with the armed forces.[20] Whereas the constitutional prerogatives and institutional arrangements of control have been something of an evolving muddle, oversight as exercised by Parliament has been a comparative success story. Several committees have played a significant part in the process of oversight and have steadily improved their performance in the oversight function. The National Defence Committee (*Komisja Obrony Narodowej*) has played the central role in parliamentary oversight.[21] The range of issues covered by the defence committee is comparable to those in long-standing NATO member states. Other institutions also have oversight functions. The Constitutional Tribunal (*Trybunal Konstytucyny*), the Supreme Chamber of Control [audit office] (*Najwyższy Izba Kontroli*) and the Citizen's Ombudsman (*Rzecznik Praw Obywatelskich*) are three major examples of organisations that exercise in a general sense, oversight of the armed forces.[22]

The development of constitutional legal framework and their attendant institutions in the civil–military relations area has been undoubtedly a complicated process. In terms of executive control of the armed forces the path leading to the establishment of clear legal structures and institutional arrangements has certainly been tortuous. Although it must be recognised that institutions charged with oversight have had a much smoother postcommunist evolution. In the end, constitutional-legal-institutional development has resulted in changes that not only reflect a maturing process of democratic control of the armed forces but also consolidate it within a solid framework of legal norms and institutions.

Professionalisation and Military Reform

Since the early 1990s, professionalisation is an issue that has been intimately bound up with all aspects of the transformation of Poland's armed forces. In their evaluation of 'professionalisation', Forster, Cottey and Edmunds argue that a normative definition of professional armed forces means that they 'accept that there role is to fulfil the demands of the civilian government of the state and are capable of undertaking military activities in an effective and efficient way'.[23] The starting point of the discussion of professionalisation considers the 'roles' postulated for the Polish armed forces in Poland's postcommunist security and defence policy.

Security and Defence Policy: Impact of Goals

In response to the changing conditions of the post Cold War strategic environment, Poland 's 'defence' or 'military' doctrine has evolved in line with the changing conditions. In content, a significant watershed regarding the evolution of doctrine was Poland's entry into NATO in March 1999. NATO membership and the changed international security environment post 11 September continue to shape the evolution of security and defence 'doctrine'. The Polish use of the term 'doctrine' encompasses a wide range of levels but, like its British counterparts, it represents a 'body of thought which underpins the development of defence policy'.[24] In 1990 and again, in 1992, the Polish government published 'doctrinal' texts setting out the first post Cold War view of potential threats and the new purposes and tasks of the armed forces. The 'Defence Doctrine of the Polish Republic' was adopted in Spring 1990 and its replacement, the 'Security Policy and Defence Strategy of the Republic of Poland', officially was accepted in November 1992. The former was clearly a first effort revision that still reflected some of the assumptions of the former communist regime, while the latter document more definitively met the security desiderata of the new political order. Following Poland's entry into NATO, the 1992 'Security Policy and Defence Strategy of the Republic of Poland' was replaced by two complementary documents that separated discussion of security and defence policy: 'Security Strategy of the Republic of Poland' (4 January 2000) and 'The National Defence Strategy of the Republic of Poland' (23 May 2000). The most recent document, 'The National Security Strategy of the Republic of Poland', emerged in September 2003 and takes into account the post 2001 international security environment.

The 1990 statement of doctrine still saw as an 'important element' Poland's 'membership in the Warsaw Pact'; the 1992 document made clear that 'Poland is striving towards NATO membership' (and the EU/WEU) as the central goal of its security policy.[25] Despite the fundamental shift in goals, there were significant elements in continuity between the two documents; both of them underscored the reassertion of national sovereignty that lay at the apex of defence doctrine. 'The strategic defence goal of the Republic of Poland', stated the 1992 document, 'is to uphold the nation's sovereignty, independence, and territorial inviolability'.[26] It also envisaged a secondary purpose in which the armed forces operated in coalition with other allied states either abroad or on Polish soil in support of international security.[27] The balance in the early 1990s was clearly tilted toward territorial defence although the way was left open for greater involvement in power projection.

The 'Security Strategy of the Republic of Poland' adopted in January 2000 marked a major shift away from the emphasis on defence of national territory.[28] Moreover, unlike the earlier statement of security policy in 1992, Poland could now place its security firmly in the context of NATO. The 'Security Strategy' indicated that 'Poland's priority is for the Atlantic Alliance to maintain its capacity to perform its functions as an effective organisation of collective defence and to ensure reliable allied solidarity'.[29] Although Poland's national security must now be viewed as firmly within the embrace of NATO, in terms of the tasks envisaged for the armed forces, defence of national territory now sits alongside a growing requirement to project military power:

> Operating both within the national defence system and within the NATO system, the Armed Forces of the Republic of Poland are ready to carry out three kinds of strategic tasks: defence-related tasks in the event of war (repelling a direct aggression against the territory of Poland or participation in repelling aggression against another allied State), crises-management (also within the framework of missions run by international organisations) and stability-enhancing and conflict-prevention tasks in peacetime.[30]

The 'Security Strategy' made specific mention of the impact that these 'strategic tasks' would have on the development of the Polish armed forces. Unlike earlier incarnations of security policy, there was far more emphasis on capability to participate in 'crisis-management operations outside Polish territory'.[31] In addition to traditional territorial defence roles, the armed forces now acquired potentially more distant operational requirements:

The operational forces are chiefly prepared for action within the framework of allied, multinational formations. Their size, level of preparedness, ability to regroup and conduct protracted operations with a minimum of casualties will be consistent with their obligations to mount together with the allied forces common defence operations and crisis-response operations, including those outside Polish territory and at considerable distance from their bases. Modern equipment, mobility and considerable operational versatility will characterise those forces.[32]

With the publication of the 'The National Security Strategy of the Republic of Poland' in September 2003 any lingering pretence of a duality of operational roles between traditional territorial defence and external projection of military power effectively disappeared. Although the 'safeguarding of Poland's sovereignty and independence, border inviolability and territorial integrity' remained the 'fundamental security policy objectives', it was recognised that the 'line of distinction between the external and internal security aspects' had become less distinct.[33] The 2003 National Security Strategy goes on to emphasise the growing 'importance of the international factor'.[34] The consequence of the reframing of Polish security doctrine in the wake of the emergence of the 'war on terror' has been a decisive shift in the roles and tasks of the Polish armed forces. The traditional territorial defence model is no longer seen as relevant: 'As the nature of the security threats evolves, static armed forces designed for territorial defence will be gradually phased out in favour of advanced, mobile, highly specialised units.'[35] In the prevailing conditions of international security, the Polish 'Armed Forces act together to enforce security in situations of threats of terrorist attacks on Poland's territory and take part in NATO's area and out of area operations as part of the antiterrorist coalition'.[36]

The evolution of Polish security and defence doctrine policy and the policy since the collapse of communism has been shaped by the objective of joining the Atlantic Alliance and a gradual shift away from armed forces suitable for static territorial defence toward a more mobile and deployable military force. Entry into NATO, has, if anything, accelerated the process of transformation of security and defence doctrine from a narrow traditional national defence focus to one that is more outward looking, viewing the provision of national security to be the function of a more 'internationalist' role for the armed forces. While providing a clear framework of goals for determining the role and function of the armed forces, these goals have had a major influence on the reform and future directions of the capabilities and force structures of the Polish armed services.

Development of Force Structure and Capabilities

The dramatic scope of the changes to the Polish armed forces stemming from policy changes is illustrated by reductions in manpower. In 1988, the manpower of the Polish armed forces numbered just over 400,000 men. Two years later, this total had been slashed by a quarter. Between 1991 and 1995, manpower totals stabilised in the vicinity of 280,000 men although there continued to be a perceptible reduction year on year. By March 1997 a 'definitive' manpower model for the Polish armed forces emerged that entailed even reductions to the level of 180,000 by 2003.[37] However, discussion did not end there, and policy-makers considered still deeper manpower cuts. By the new millennium, the manpower ceiling for the Polish armed forces was targeted to be 150,000 with the financial resources thus gained through manpower reductions directed toward modernisation of the armed forces.[38] Discussion of future plans for the armed forces suggest that further deeper cuts in manpower may lie ahead coupled to ambitious modernisation plans.[39]

The revision of Polish security and defence policy in 1992 triggered a series of changes to the force structure and capabilities of the Polish armed forces that have continued to evolve with each major development of policy. In the eight years following the demise of communism in 1989, most of these changes had the quality of dismantling something that existed before rather than building something new. As a consequence, the changes that occurred to the armed forces developed in a piecemeal fashion. General Henryk Szumski, the recently retired Polish CGS confirmed this lack of a coherent approach to reform in this period: 'Our army has been in the process of reforms for many years now. Necessary as they were, those reforms were superficial, partial, and not based on a final vision. Separate segments were sorted out, while having in mind no complete picture of how the Army should look in the future.'[40]

By 1997, however, a coherent reform plan emerged called 'Tenets for the Programme of the Armed Forces Modernisation, 1998–2012'.[41] The 'Plan 2012', as it became known, was adopted by the postcommunist government of the day as official policy. Plan 2012 still remained policy after parliamentary elections in autumn 1997 returned a new government made up of former Solidarity parties. Although the new government did not discard Plan 2012 it decided to review it and make necessary 'corrections'.[42] The most important correction it made was to take into account the requirements for integrating Poland into NATO.[43] Accordingly, the modernisation plan emerged with a new title: 'Programme for Integration into the North Atlantic Treaty Organisation and Modernisation of the Polish Armed Forces 1998–2012'. The alterations to the plan appeared to have been minimal and the modified version was confirmed as official policy.[44] Although Plan 2012 went a long way in reshaping the force structure of the Polish armed services, in terms of the size of the forces planned it remained too optimistic particularly when viewed from the perspective of the yawning resource gap. The design of Plan 2012 was not only too ambitious in terms of resources but also grounded in territorial defence mindset. Bolder thinking would have to await Polish membership in NATO.

Poland's entry into NATO and fuller integration into the Alliance's planning cycle prompted further changes to the long-term plans for reforming the Polish armed forces. By autumn 2000, Plan 2012 was formally superseded by a 'Programme for Modernisation of the Armed Forces 2001–2006'.[45] The revised programme placed the reform of the armed forces firmly within the NATO planning cycle and took into account Poland's contribution to the Alliance. Moreover, the revised plan took a much more realistic view of the Polish state's ability to provide resources. Under the new plan, the manpower level of the Polish forces was to be slashed to 150,000 by the end of December 2003. Obsolete equipment was to be withdrawn from service very rapidly so as not to waste resources on material of little combat value. The aim of the 'Programme for Modernisation' by trimming redundant manpower and obsolete equipment was to shift resources out of vegetative expenditure and redirect these resources to investment in new equipment.[46]

The 'Programme for Modernisation' for the armed forces reshapes the force structure in line with wider European trends toward leaner and better-equipped formations. For the Polish land forces, the number of divisions will be cut from the six of Plan 2012 to just four in the new programme. The number of brigades will drop from 23 to 18 composed of eight mechanised, one coast defence (effectively mechanised), six armoured and three light ones: air cavalry, air assault and mountain. Manpower will drop from the previously planned 107,000 to 89,000.[47] The Polish air force would roughly match its previously planned aircraft numbers on seven permanent bases, but with its manpower down by a further 7,000 to a total of 31,000.[48] For the Polish Navy, the 'Programme for Modernisation' will see its manpower reduced to 13,500 and older ships disposed of at a faster rate.[49] The 'Programme for Modernisation' in most respects is a more radical version of the previous Plan 2012 and takes better into account the shift in NATO toward more flexible and mobile forces.

With these changes to force structure and the addition of new capabilities, the operational forces of the Polish armed services will undoubtedly see an improvement in their ability to project military power outside Poland. In an effort to acquire more up-to-date equipment and become more interoperable with its German neighbours in NATO's Allied Rapid Reaction Corps (ARRC), the Polish Defence Ministry purchased an equipment package for the Polish 10th Armoured Cavalry Brigade that included 128 Leopard 2A4 Main Battle Tanks.[50] The army's other procurement priorities include Mi-24W Hind helicopters HMMMV (Hummer) terrain vehicles and acquisition of the Patria AMV wheeled armoured APCs.[51] The Polish Navy is currently operating a small logistic support vessel with plans to acquire larger multirole logistic ships that could support a proposed marine battalion.[52] The acquisition of Oliver Hazard Perry class frigates from the United States has given the Polish Navy a blue water capability beyond the Baltic Sea.[53] For the Polish air force, the decision to purchase 48 Lockheed Martin F-16C/D Block 52M+ multirole fighters means a significant enhancement of capability. The modernisation and enhancement of airlift capacity is also underway with 8 C-295M light tackle transport acquired new from its Spanish manufacturer while 6 used C-130 transport aircraft are to be acquired from the United States.[54] The priorities exhibited in equipment purchases point to the acquisition of capabilities useful for participation in coalition operations outside of Poland.

Toward All-volunteer Professional Armed Forces

Poland, like so many countries after the Cold War, has to face the question of whether it should make its armed forces all volunteer in composition. In the context of the Polish debate on professionalisation, full-time career or fixed term contract service personnel are considered 'professionals' but conscripts do not fall into the professional category. It is this narrow view of professionalisation that prevails in Polish military thinking. In 1998 General Józef Buczyński, head of the Polish Ministry of Defence personnel department, noted in an interview that the emphasis on all-volunteer forces is increasingly a military necessity: 'In the face of the dynamic development of military technology, military service is becoming a domain of professionals. In future, there will be no room in it for amateurs and undereducated soldiers.'[55]

The 'National Defence Strategy of the Republic of Poland' published in May 2000 indicated that 'operational units shall have the level of professionalisation of 50%'.[56] The 'Programme for Modernisation of the Armed Forces 2001–2006' anticipates that 75,000 will be 'professional' service personnel consisting of 22,500 officers, 22,500 warrant officers and 30,000 NCOs.[57] For the Polish Navy, the increase in the proportion of professional service personnel will be between 55% and 60%.[58] Even in the army, however, some units will have a much higher proportion of 'professional' soldiers. Clearly in the case of units most likely to be deployed operationally, the trend is to see higher proportions of 'professional' service personnel grow at the expense of conscripts.

Since Poland regained its independence in 1918, conscription has been a central feature of its military manpower provision. In autumn 1990, the length of service of conscripts was reduced from twenty-four months to the current eighteen months. A further reduction of length of service to twelve months was put in place in January 1999.[59] After 2004, a further reduction to nine months is underway.[60] The reduction in the length of military service certainly accords well with changing public perceptions regarding national service. Certainly the armed forces has been disappointed for some time by both the physical and mental abilities of its conscripts.[61] The frequent and well-publicised examples of the bullying of conscripts have not added to public popularity of national service. Despite military efforts to crack down on bullying, it may be seen as one of the factors contributing to public disenchantment with conscription. Opinion polling by the Military Sociological Institute suggests that fewer young people want to serve in the armed forces and that many people view national service as a waste of time.[62]

From the point of view of both the Polish public and the military, the less attractive qualities of conscripts and conscription may only be fully resolved by all-volunteer armed forces. In May 1998, an opinion poll conducted by the CBOS organisation indicated that some two-thirds of those consulted support an all-professional (all-volunteer) army without conscripts.[63] It is clear that the Polish Ministry of National Defence is moving toward all-volunteer professional armed forces. The introduction of new legislation to create a professional enlisted corps to be in place by July 2004 is an important step to creating all-volunteer armed forces.[64] The Polish Defence Minister, Jerzy Szmajdziński recently indicated that by 2010, a full 70% of personnel would be all-volunteer professionals.[65] The creation of all-volunteer Polish armed forces seems only a question of not whether but when.

The Military and Society

Polish society has had a long-standing *affaire d'amour* with its armed forces. Rooted in the historical experiences of the nation in the nineteenth and twentieth centuries, the popularity of the military is woven into the fabric of national consciousness.[66] In broad terms, this support for the armed forces remains true today in Polish society although society's approval cannot be viewed as uncritical. This can be illustrated by a series of opinion polls assessing public views of the armed forces. An opinion poll by the Social Research Laboratory (PBS) that sought a positive or negative opinion of institutions in June 1999 also ranked the armed forces in seventh place but with only 41% of respondents having a positive opinion and 59% taking a negative view.[67] As if to contradict the sliding level of positive opinion of the armed forces as an institution, polling evidence on the Polish public's trust of the armed forces as an institution has been consistently very high for a long period of time. For example, in February 1998 and April 2002 the armed forces occupied first place among institutions at 71 and 79% of respondents.[68] What this opinion polling evidence tells us about attitudes in Polish society is that as an institution, the armed forces are generally held in high esteem but it is one that may have its problems and does not in every respect function to the Polish public's liking.

The values of Polish society condition the relations with the armed forces but in ways generally not consistent with the 'postmodern' agenda found in the societies of long-standing NATO member states.[69] Indeed, Polish society may be characterised as being in a contradictory position where traditional values still run strongly in the mainstream of Polish society but are increasingly challenged by processes of major economic and social change driven by the postcommunist transformation.[70] Although Poland is counted among the more successful examples of economic transformation, the cost to society has been impoverishment and increasing economic stratification.[71] The contradictory picture is best illustrated by contrasting the fact that Poland since the end of communism has seen the expansion of the number of Roman Catholic archdioceses and dioceses that suggests the strength and enduring qualities of traditional values. At the same time, however, the country has seen growing opportunities for women in business and the professions.[72]

On the issue of women in the armed forces, more traditional patterns and attitudes are deeply entrenched. Women are clearly one group with extremely modest representation in the armed forces. Apart from the impact of traditional values in Polish society, there exist major barriers to increasing the number of women in the armed forces in the military itself. In January 2001, only 277 women served in the Polish armed forces representing about 0.1% of the total serving personnel.[73] In comparison with other NATO members this was less that the Czech Republic (3.7%) and Hungary (9.6%) and very distant from the United States' 14%.[74] By the end of 2001, numbers of women in military service had risen to 288 with 230 in training.[75] For the time being, women represent a tiny element in the personnel structure of the Polish armed forces. Polish society, however, is changing rapidly and the traditional roles of women are coming under pressure. Over time it is inconceivable that the armed forces of Poland will be immune from the wider changes in society that are expanding the role of women. At present, the armed forces are not prepared either in attitude or in meeting the practical challenges of having larger numbers of women in the military.[76] The issue of women in the armed forces is indicative of the collision between traditional values and emerging agendas in a rapidly changing society that is bound to change the nature of relations between the armed forces and society.

Conclusion: Influences on Changing Civil–military Relations

International and Transnational Influences

The dominant external influences on civil–military relations in Poland come from the broad integration process into western institutions. The impact of this process is more likely to come from the direction of the economic and social change prompted by Poland's efforts to join the European Union (EU). The economic and social spheres have a much more generalised effect on the attitudes of Polish society. In the longer term it may lead to increasing convergence with the postmodern values of western European states. In the short to medium term, however, it is the economic and social costs of transformation and preparation for EU membership that has the most immediate impact. The competing economic and social desires in Polish society are going to make defence a lesser priority. For Poland, like many of its regional partners, 'in the absence of any direct external military threat, the internal crisis [of transformation] of each country is, by far, the dominant source of anxiety'.[77] Since joining NATO, preoccupation with economic and social transformation issues is complemented by the fact that Poles also believe that their country is safer from external threats.[78] Although NATO may bring to Polish society a feeling of security, it also presents some challenges to the military–society relationship.

The shift of the armed forces role toward more expeditionary operations means that Polish society will be confronted with deployments to conflicts outside national borders that carry a number of risks not the least the possibility of Polish casualties. Members of the Polish armed forces are willing enough to serve abroad as military opinion polling indicates.[79] The support of Polish society, however, cannot be taken so easily for granted. When examining the Polish public's reaction to sending troops to join coalition operations in Afghanistan in October 2001, 65% were opposed to sending troops.[80] By January 2002, opposition considerably lessened but views were evenly split with 43% in favour and against.[81] This suggests that the support for Polish participation in operations abroad, whether or not in the context of NATO, will be given on a case by case basis. This mixed picture is tested most seriously with one of Poland's largest deployments abroad—Iraq.

In August 2003, Poland deployed around 2,500 troops in a brigade-sized deployment that formed part of a Polish led multinational division controlling a central zone in Iraq. In military terms, it is certainly the most substantial and ambitious deployment of Polish forces abroad since the Second World War. Moreover, as a catalyst for accelerating the transformation of the armed forces and enhancing professionalism, it has undoubtedly had a positive impact.[82] The effort clearly enjoys substantial American support and indirectly that of NATO. As in the case of the earlier deployment to Afghanistan, Polish public support cannot be taken for granted. Indeed, Polish public opinion had serious misgivings about embarking on the operation and the loss of a Polish officer in November 2003 did not enhance public support.[83] What the developments indicate is that the Iraq deployment indicates that a national debate on more expeditionary roles for the armed forces is only beginning.[84] As for Polish forces in Iraq, they are facing their most serious challenge. With Polish public opinion, at best, divided on the Iraq deployment, the outbreak of the Shia uprising in April 2004 will provide the most serious test of the professionalism of the armed forces and the Polish public's support of the Iraq operation.[85]

Domestic Influences

The domestic influences on the development of civil–military relations in Poland is likely to be influenced by a number factors into the foreseeable future. These include the move toward all-volunteer professional armed forces, the unpopularity of conscription in Polish society and economic constraints on defence spending. The new missions for the Polish armed forces underscored by NATO membership have contributed to the emerging prospect that Poland will eventually have all-volunteer professional armed forces. The ultimate consequence of any move toward all-volunteer professional forces would be an end to conscription in Polish society. Although conscription is impacting on fewer individuals, should it disappear entirely, it would sever a long established link between the armed forces and society. It is a link, however, that Polish society seems all too willing to break.[86]

In the next decade (as in the past one), the most important challenge in the military–society relations is the issue of funding. The Polish Defence Minister, Szmajdziński stated that 'the greatest problem of the Polish military is its chronic underfunding'.[87] The significant changes to the employment of the armed forces entailed by NATO, professionalisation and the end of conscription require the application of substantial resources. With the competing resource demands of a difficult economic and social transition, Polish society is less willing to make defence a priority budget item. As is often the case on key issues at the centre of military–society relations, democratic societies find it easier to will the ends rather than the means of achieving them.

Notes

[1] The opinions expressed in this chapter are those of the author and do not necessarily reflect those of either RMA Sandhurst or the UK Ministry of Defence.
[2] A. Żebrowski, *Kontrola cywilna nad siłami zbrojnymi Rzeczpospolitej Polskiej*, (Warsaw: Bellona, 1998), pp. 43–48.
[3] Law on the Duty to Defend the Republic of Poland, Supplement to the Proclamation of 11 December 1991 of the Minister of National Defence, [Amendment of] Law of 21 November 1967 on the General Duty of Defending the Polish People's Republic', *Dziennik Ustaw*, No. 4, 22 January 1992. Item no. 16 in: JPRS-EER-92-111-S, 20 August 1992.
[4] All references to the Small Constitution are from the following: *The Rebirth of Democracy: 12 Constitutions of Central and Eastern Europe* (Council of Europe Press, 1995), pp. 381–426.
[5] See Article 26 (2) in: *The Rebirth of Democracy*.
[6] See Article 146 (7), (8) and (11) in: *Konstytucja Rzeczypospolitej Polskiej 1997*.
[7] See Article 134 in: *Konstytucja Rzeczypospolitej Polskiej 1997*.
[8] See Articles 140, 141 and 144 in: *Konstytucja Rzeczypospolitej Polskiej 1997*.
[9] 'Law on the Duty to Defend the Republic of Poland, Supplement to the Proclamation of 11 December 1991 of the Minister of National Defence.
[10] Tomasz Niewiadomski, 'Jest KOK czy go nie ma', *Rzeczpospolita*, 7 July 1999.
[11] Niewiadomski, 'Jest KOK czy go nie ma'.
[12] S. Koziej, *Kierowanie obroną narodową Rzeczpospolitej Polskiej* (Warsaw: DBM Paper no. 37, 1996), pp. 38–67.
[13] Żebrowski, *Kontrola cywilna nad siłami zbrojnymi Rzeczpospolitej Polskiej*, pp. 114–115.
[14] See Article 135 in *Konstytucja Rzeczypospolitej Polskiej 1997*.
[15] See report by the Polish News Agency PAP, 20 January 1998.
[16] For a comprehensive view of the status of KOK see the following: Niewiadomski, 'Jest KOK czy go nie ma'.
[17] Stanisław Jarmoszko, *Wojsko Polskie: pierwszej dekady transformacji* (Toruń: Adam Marszałek, 2003), p. 41.
[18] Andrew Michta, *The Soldier-Citizen: The Politics of the Polish Army after Communism* (Basingstoke: Macmillan, 1997), pp. 87–91, and *Janusz Onyszkiewicz ze szczytów do NATO* (Warsaw: Bellona, 1999), pp. 203–204.

[19] 'Ustawa z 14 grudnia 1995 r. o urzędzie ministra obrony narodowej', *Rzeczypospolita*, 1 February 1996.
[20] See Small Constitution in: *The Rebirth of Democracy: 12 Constitutions of Central and Eastern Europe* (Council of Europe Press, 1995), pp. 381-426 and *Konstytucja Rzeczypospolitej Polskiej 1997*.
[21] Żebrowski, *Kontrola cywilna nad siłami zbrojnymi Rzeczpospolitej Polskiej*, pp. 88-102.
[22] Żebrowski, *Kontrola cywilna nad siłami zbrojnymi Rzeczpospolitej Polskiej*, pp. 154-224.
[23] Anthony Forster, Andrew Cottey & Tim Edmunds, 'Introduction: Professionalisation of Armed Forces in Post Communist Europe', in: Anthony Forster, Tim Edmunds & Andrew Cottey (eds), *The Challenge of Military Reform in Postcommunist Europe: Building Professional Armed Forces* (Houndmills: Palgrave, 2002), p. 6.
[24] *British Defence Doctrine*, Joint Warfare Publication (JWP) 0-01 (London: HMSO, 1997), p.1.2.
[25] 'Defence Doctrine of the Polish Republic', 21 February 1990, in: *Żołnierz Wolności*, 26 February 1990 in: JPRS-EER-90-038.
[26] 'Security Policy and Defence Strategy of the Republic of Poland', 2 November 1992 in: *Wojsko Polskie: Informator '95* (Warsaw: Bellona, 1995), pp. 16-32.
[27] Ibid.
[28] 'Security Strategy of the Republic of Poland', 4 January 2000, web site of the Polish Ministry of Foreign Affairs, http://www.msz.gov.pl/english/polzagr/security/index.html
[29] Ibid.
[30] Ibid.
[31] Ibid.
[32] 'Security Strategy of the Republic of Poland', 4 January 2000.
[33] 'The National Security Strategy of the Republic of Poland', September 2003, web site of the Polish Ministry of Foreign Affairs, http://www.msz.gov.pl/
[34] Ibid.
[35] Ibid.
[36] Ibid.
[37] Paweł Wronski, 'KSORM on Changes in Armed Forces', *Gazeta Wyborcza*, 14 March 1997 in: FBIS-EEU-97-073.
[38] 'Onyszkiewicz: 150-tysięczna armia', Polish News Agency PAP, 23 May 2000. See also the following: Interview of the Polish Defence Minister Janusz Onyszkiewicz, 'Kłopoyliwe zmiany na korzyść', *Polska Zbrojna*, 16 April 2000, p. 27 and Interview with the Chairman of the Parliamentary Defence Committee, Bronisław Komorowski, Warsaw Radia, 27 December 1999, in: FBIS-EEU-1999-1227. For earlier developments see: Testimony of the Janusz Onyszkiewicz, Minister of National Defence, to the Polish Parliamentary Defence Committee, 5 May 1998, Maj. Ryszard Choroszy, 'Przyspieszenie', *Polska Zbrojna*, 29 May 1998 and 'Onyszkiewicz: Mniej wojska, ale nie kosztem zdolności bojowych', Report of Polish news agency PAP, 12 December 1999.
[39] Mieczysław Cieniuch, 'Armia drugiej dekady', *Polska Zbrojna*, 10 August 2003.
[40] Interview of Gen Henryk Szumski, *Polska Zbrojna*, 2 May 1997.
[41] See for example: 'Armia 2012', *Polska Zbrojna*, 12 September 1997, Wojciech Luczak, 'Wojsko XXI wieku', *Zycie Warszawy*, 10 September 1997, Z.L., 'Armia mniejsza i silniejsza', *Rzeczpospolita*, 10 September 1997, and Paweł Wronski, 'Armia XXI wieku, *Gazeta Wyborcza*, 10 September 1997.
[42] 'Armia 2012 do korekty (?)', *Polska Zbrojna*, 26 December 1997, 'Armia 2012 do poprawki', *Polska Zbrojna*, 6 March 1998, and 'Armia 2012 poprawiona', *Polska Zbrojna*, 12 June 1998.
[43] Janusz Zemke, 'Urodzaj na programy', *Polska Zbrojna*, 13 March 1998.
[44] Report of Polish news agency PAP, 13 July 1998.
[45] 'Program modernizacji sił zbrojnych 2001-2006', September 2000, Polish Ministry of Defence web site, http://www.wp.mil.pl/aktualnosci/5_609.html

[46] See interview of the Polish Defence Minister, Bronisław Komorowski, 'Wystartujmy z planem 6-letnim', *Polska Zbrojna*, 30 June 2000 and article by Zbigniew Lentowicz, *Rzeczypospolita*, 15 June 2000.
[47] Janusz B. Grochowski, 'Armia 2006', *Polska Zbrojna*, 1 April 2001.
[48] 'Prostowanie skrzydeł', *Rzeczpospolita*, 19 April 2001.
[49] Artur Goławski, 'Cała naprzód!', *Polska Zbrojna*, 8 April 2001.
[50] Grzegorz Holdanowicz, 'Polish Leopard 2 delivery a major NATO milestone', *Jane's Defence Weekly*, 25 September 2002.
[51] Christopher F. Foss, 'Patria'a armoured vehicle for Poland takes shape', *Jane's Defence Weekly*, 17 September 2003.
[52] Grzegorz Holdanowicz, 'Poland plans logistics ships and marine force', *Jane's Defence Weekly*, 18 September 2002.
[53] Paweł Wroński, 'Nasza niemłoda fregata', *Gazeta Wyborcza*, 14 March 2000.
[54] Grzegorz Holdanowicz, 'First new transport aircraft arrives for Poland's air force', *Jane's Defence Weekly*, 27 August 2003 and Grzegorz Holdanowicz, 'Poland to acquire C-130s', *Jane's Defence Weekly*, 25 June 2003.
[55] Interview General Józef Buczyński, *Polska Zbrojna*, 19 June 1998.
[56] 'The National Defence Strategy of the Republic of Poland', 23 May 2000 (Warsaw: Ministry of National Defence, 2000).
[57] 'Program modernizacji sił zbrojnych 2001–2006', September 2000.
[58] Grochowski, 'Armia 2006'.
[59] Report Polish news agency PAP, 3 November 1998.
[60] Jerzy Szmajdziński, 'Transformacja', *Polska Zbrojna*, 21 March 2004.
[61] Tadeusz Mitek, 'Słabsze pokolenie', *Polska Zbrojna*, 10 March 2000.
[62] 'The Armed Forces, A Waste of Time?', *Gazeta Wyborcza*, 28–29 May 1997 in: FBIS-EEU-97-153.
[63] Renata Wróbel, 'Obowiązek, którego łatwo uniknąć', *Rzeczpospolita*, 22 May 1998.
[64] Jerzy Szmajdzinski, 'Transformacja', *Polska Zbrojna*, 21 March 2004.
[65] 'Armia bardziej zawodowa', *Rzeczypospolita*, 1 April 2004.
[66] Jerzy J. Wiatr, 'The Public Image of the Polish Military: Past and Present', in: Catherine McArdle Kelleher (ed.), *Political–Military Systems: Comparative Perspectives* (Beverly Hills: Sage Publications, 1974), pp. 199–201.
[67] Opinion poll by PBS, *Rzeczpospolita*, 29 June 1999.
[68] Opinion polls by OBOP, in: Polish News Agency PAP, 16 February 1998 and 'Polacy ufają mundurowi', *Polska Zbrojna*, 28 April 2002.
[69] For a detailed study of the attributes of the postmodern military see: Charles C. Moskos, John Allen Williams & David R. Segal (eds), *The Postmodern Military: Armed Forces after the Cold War* (New York: Oxford University Press, 2000).
[70] Marek Ziółkowski, 'The Pragmatic Shift in Polish Social Consciousness: With or Against the Tide of Rising Post-Materialism?', in: Edmund Wnuk-Lipiński (ed.), *After Communism: A Multidisciplinary Approach to Radical Social Change* (Warsaw: ISP-PAN, 1995), p. 170.
[71] Jan Danecki, 'Social Costs of System Transformation in Poland', in: Stein Ringen & Claire Wallace (eds), *Societies in Transition: East-Central Europe Today: Prague Papers on Social Responses to Transformation* Vol. I (Prague: CEU Press, April 1993), pp. 47–60.
[72] Jan Kofman & Wojciech Roszkowski, *Transformacja i postkomunizm*, (Warszawa: ISP-PAN, 1999), p. 101 and p. 125.
[73] *National Report—Military Service of Women in Poland—2001*, Polish Ministry of National Defence website, URL: http://www.wp.mil.pl/_en_about/e_6_n.htm (accessed 27 August 2001).
[74] Anna Dąbrowska, 'Od makijażu do kamuflażu', *Polska Zbrojna*, 9 December 2001, p. 14.
[75] Ibid.
[76] Report of Polish news agency PAP, 23 November 1998.
[77] M. Boguszakova, I. Gabal, E. Hann, P. Starzynski & E. Taracova, 'Public Attitudes in Four Central European Countries', in: R. Smoke (ed.), *Perceptions of Security: Public Opinion and*

Expert Assessments in Europe's New Democracies, (Manchester: Manchester University Press, 1996), p. 34.

[78] Opinion poll by SMG/KRC, *Zycie Warszawy*, 27–28 March 1999.

[79] A series of opinion polls by the Military Sociology Research Organisation indicated that about 80% of professional soldiers supported deployments outside Poland, *Polska Zbrojna*, 26 August 2001, p. 8.

[80] Opinion poll in *Rzeczpospolita*, 27–28 October 2001.

[81] Opinion poll by CBOS, *Polska Zbrojna*, 10 February 2002, p. 8.

[82] Interviews with officials from the Polish Ministry of National Defence, October 2003.

[83] 'Nie dla misja', *Rzeczpospolita*, 28 November 2003 and 'Rośnie poparcie dla irackiej operacji', *Rzeczpospolita*, 25 September 2003.

[84] Interviews with officials from the Polish Ministry of National Defence, October 2003.

[85] Monika Scislowska, 'Poland Said Committed to Iraq War', *Guardian Unlimited*, 7 April 2004, Guardian Unlimited website. Available at http://www.guardian.co.uk (accessed 7 April 2004).

[86] Paul Latawski, 'Professionalisation of the Polish Armed Forces: "No Room for Amateurs and Undereducated Soldiers"', in: Forster, Edmunds & Cottey (eds), *The Challenge of Military Reform in Postcommunist Europe: Building Professional Armed Forces*, p. 28.

[87] Article by Polish Defence Minister, J. Szmajdzinski, *Polska Zbrojna*, 25 April 2002.

Democracy and Defence in Latvia

Thirteen Years of Development: 1991–2004

JAN ARVEDS TRAPANS

At NATO's November 2002 Prague Summit, Latvia was one of the countries which was judged to be a suitable member of the Alliance, both militarily and politically. In April 2004 it acceded to full membership of NATO. Twelve years earlier, when the Soviet Army withdrew from the country '(all) that was left behind consisted of 26 sunken submarines and ships leaking acid, oil, and phosphorous. On this foundation Latvia began building its armed forces.'[1] The political structure was not in such a dire condition, but great efforts were needed to establish democracy. Latvia had a Supreme Soviet, which was to be replaced by a *Saeima*, or Parliament; it had a Soviet Constitution, which also had to be replaced: there was no Foreign Ministry and no Defence Ministry. Western observers said of Latvia: 'It is worth recalling that, at the beginning, (no) national Armed Forces existed; that the military infrastructure was in ruins; that equipment and logistical support were almost non-existent; that public support for the professional military was low; that training and experience ... had been gained in a very different Soviet system.'[2] Latvia has had thirteen years to establish a democratic system, its armed forces, and democratic control over the armed forces.

Democratic Control of Armed Forces

In principle, the fundamentals of democratic control need to exist in three areas. First, the military can have no influence in domestic politics in order to build safeguards for its powers and privileges. Second, it cannot exert influence over policy so as to shape the defence posture of the country to its liking. Third, it has to accept democratic control over defence policy. Governments propose and parliaments confirm the defence budget, procurement, and related matters. The soldiers should be 'the apolitical servant[s] of the democratic government ... [they] should confine [themselves] to implementing decisions made by those authorities'.[3] The principles are found in Western societies, which have had centuries to consolidate their civil–military relationships; Latvia has had a little over a decade. It had to erect, simultaneously, a political establishment, a military establishment, have both parts understand what democratic control means, and act accordingly.

The basis for Latvian civil–military relations is found in the Latvian Constitution. Latvia restored the Constitution of 1922, with some minor amendments. In retrospect, the decision was a felicitous one. The Constitution has provided a solid foundation for democratic control and Latvia avoided the frequently encountered constitutional problem that has plagued many former socialist states. Namely, imprecise constitutional provisions about the civilian structure of control over the armed forces have led to conflicting interpretations by Presidents and Prime Ministers about their political prerogatives. This has caused political contests or rivalries between the President, the Government (i.e. the Defence or Prime Ministers), and the Parliament. Latvia's Constitution gives a clear delineation of political powers and responsibilities. The President, as the Head of State is the nominal Commander of the armed forces in peacetime and appoints a military wartime Commander. The President can declare war only following a decision by the Parliament but can initiate necessary defence measures in the event of aggression without consultation. The Parliament or *Saeima* approves the budget, settles on the size of the armed forces, confirms the commander of the national forces, and decides on Latvia's participation in international missions. The parliamentary committees for security and defence affairs consider all pertinent legislation. Latvia's constitutional and legal framework is firm and clear. Representatives of NATO have not found anything that needs change or improvement.

Constitutional provisions had to be complemented by additional, specific laws, because Latvia was building its armed forces *ab initio*. Moreover, new requirements arose, like sending units abroad for peacekeeping missions. Some laws, written by inexperienced parliamentarians, were ambiguous, imprecise, and revisions were needed. Currently, there are eight relevant laws, and the National Security Law of 2001 determines the composition and tasks of the national security system, the competence of the persons or institutions responsible for the national security system and the principles and procedures of coordination, implementation and control of their activities.[4] Service personnel have accepted the essential principle of democratic civilian control, a development validated through external audit of Latvian civil–military relations by the North Atlantic Parliamentary Assembly, the US Army Latvian Assessment Document (or Kievenaar Study) of 1997, and the Latvian Membership Action Plan for NATO (MAP).[5] The Defence Ministry has to respond to parliamentary demands; there must be 'transparency and accountability'. There has been some manoeuvring by the military in defence policy, mostly during the early independence years, but what has taken place falls entirely within the limits of what the Western military does. All-in-all, the military has acquitted itself as a politically neutral body, with no attempts to exploit political parties.

Constitutional and legal arrangements, no matter how well constructed, do not provide the necessary expertise on policy makers. Without competence in the Defence Ministry, the Parliament, and other parts of the government, without a civil service that can meet the military on equal ground, there can be no functioning civil–military relations. Reform needs reformers. In the new democracies there was a lack of national governmental capacity, of people with overall competence for defence policy formulation and planning. Western experts have dealt with democratic control with attention directed primarily to its constitutional and legal aspects. According to this philosophy, once the armed forces were placed under layers of control—civilian, democratic, parliamentary—a fundamental settlement would be in place. In the real world it is the capability of the civilians in Defence Ministries and Parliaments that weigh heavily in the balance.

Latvia's Defence Ministry, once established, was staffed with a small number of inexperienced civil servants, most of them quite young. It took time to develop the full range of mechanisms and procedures for the conduct of government business which the nations of Western Europe have had the opportunity to build and refine over the last half century and more. However, to its advantage, Latvia did not inherit a large, stodgy, Warsaw Pact defence bureaucracy either. Thus, there were advantages as well as disadvantages to the situation, more of the former than the latter. Western observers commented that 'enormous weight rests on the shoulders of a small group of admirable young men and women, who struggle to keep on top of the problem in hand'. However, by the late 1990s their assessment was that 'there now exists a civilian (Ministry of Defence) functioning with growing confidence and effectiveness to establish their roles and positions *vis-à-vis* the Armed Forces and their Headquarters'.[6] In the years after the comments were made, the confidence and effectiveness of the civilians has increased considerably; their number has increased only slightly. In some other new democracies, Defence Ministries are bulging with staff—much to their disadvantage.

Parliamentary committees consider legislation concerning defence and security. The Parliament approves the budget in a transparent process, decides on the size and structure of the armed forces, approves the peacetime military commander of the armed forces, and decides on participation in international missions. Civilian–military relations have not entered some blind legal ally. There have been no political conflicts among the civilians as to who actually controls the armed forces. In countries with democratic systems and free market economies, present-day defence planning is complicated and involves long-term economic analyses. It is also international in nature, overlaps with foreign policy, and, in Latvia's case, has a large Baltic, or regional dimension. Moreover, the soldiers have moved across a demarcation line which once separated military and non-military responsibilities. The tasks of a professional officer, for example, may include administrative work, preparing documents for the cabinet, involvement in developing the defence budget, or appearing before a parliamentary committee. The military, like the civilians, requires a new set of skills: political, managerial, and international. Latvia has developed a functioning civilian–military defence community.

Professionalisation and Military Reform

Latvia's armed forces emerged from the turbulent events of 1990 and 1991 as the Baltic States wrested themselves away from the collapsing Soviet empire. Armed formations were created *ad hoc* to respond to immediate needs. In 1991 Latvia's Parliament established volunteer Border Guards to demonstrate that it had control over its own borders, as a sovereign state is supposed to. A unit to defend the Parliament also was created. Self-defence forces, the National Guard or *Zemessardze* arose spontaneously throughout the countryside. (Subsequently, Latvia's volunteer militia received substantial assistance from the Michigan National Guard.) The Government called Latvian officers who had served in the Soviet Army to establish regular armed forces. A Defence Ministry was established only in November of 1991.

It took time to sort out this confusion. Considerable effort was needed to place these various formations under clear lines of civilian control. Latvia aimed to create a small regular conscript force, commanded by professional officers from the former Soviet Army. This was in addition to the National Guard, a volunteer force with commanders who saw themselves as the leaders of a 'nation in arms', but who had limited military experience. There was distrust between the small professional and large volunteer bodies. Both had their supporters. The Defence Ministry had a large volunteer militia on hand and had to make the best use of it. Resolving the complications brought about by a force structure of this type, chiefly in developing a chain of command, establishing joint headquarters, and developing training and doctrine was not easy, and many decisions had a political aspect. In time, the differences between the two military bodies have been removed. However, because there were many shortages—of weapons, equipment, facilities, and, above all, military experience—this process was slow.

Given Latvia's economic and demographic capabilities, a small conscript force backed by a large reserve component is the feasible, logical force structure. In the traditional view of defence against an external threat, two factors determine Latvia's security—geography and economy. Latvia has a large, acquisitive neighbour to the east, a sea to the west, and two friendly but small neighbours to the north and the south. Its economy is small and so is the population. Defence plans are determined by economic and demographic resources and, in Latvia's situation, there will never be sufficient resources, in terms of personnel or materiel, to counter a major external threat. As a consequence, it has had to come up with a strategy that can deny a rapid victory and make the political and economic losses to an aggressor outweigh possible gains.

These considerations have been reflected in Latvia's national security concepts, which have been revised and updated over the years. The most recent version was issued in 2002.[7] The concept says that the objectives of national security are to protect and preserve the nation's sovereignty, territorial integrity, and a democratic, constitutional form of government, market economy, national identity, and human rights. The national security of the Republic of Latvia is the ability of the state and its society to protect and ensure the country's national interests and basic values. Security also requires the maintenance of internal political stability, which is based on overall awareness of democratic development of the country and the development of a unified civil society. This in turn is based on the principle of equality of rights for all individuals. The ability of Latvia to secure its national interests depends greatly on external conditions. Latvia has joined NATO and the EU and its armed forces and civil–military relations have to meet the requirements of these organisations. Cooperation between the Baltic States is also an important feature in security. In addition, the country faces new security risks, such as organised crime and trafficking which cannot be met with traditional military means. The armed forces have three missions. The primary one is to protect the country's sovereignty and territorial integrity. In peacetime the armed forces will provide deterrence by demonstrating their readiness and capability to defend the nation's sovereignty. Second, they support the civil powers in emergency situations. Finally, under the threat of a war, or in wartime conditions, they will defend the national territory, airspace, territorial waters, and key administrative and political centres.

The Government has developed and the Parliament has approved successive national security concepts based on a threat assessment. The concepts are public documents. The national security concepts do not have the force of law but stand above party politics. In 1999, the International Defence Advisory Board to the Baltic States (a group of senior Western soldiers and statesmen) concluded that Latvia's security policy had been developed and promulgated in a democratic process.[8] The military had provided expertise but it had not exerted influence over policy decisions. The concepts present the rationale for a Western-oriented national security policy.

The concepts provided a basic statement on the missions and organisation of the armed forces and outlined a force structure. The armed forces are to consist of a conscript core backed by a large volunteer reserve. They are small, predominantly land based but with naval and air force components. Defence plans, based on the security concept, call for regular armed forces with a strength of 10,000, consisting of 4,000 volunteers and 6,000 conscripts. The reserve component would number between 35,000 and 40,000.[9] However, the defence planners understood that the envisaged force structure could not be in place for some ten years. Latvia could not afford the necessary weapons, equipment, supplies, or conduct large-scale training to build up the desired end strength in a short time. In the meantime it has adopted a strategy similar to Finland's—that of territorial and total defence. A great power aims at a swift military victory, forcing the defender to capitulate militarily and politically. A small country must deny the aggressor's objective, fighting on its home territory with extended small-scale actions. Territorial defence is a decentralised but cohesive military action. It is carried out by a small, active force, in a condition of high readiness, supported by reserve components, relatively stationary and locally mobilised, organised in defence regions and defence districts. An aggressor would be met with protracted military resistance throughout the country's territory. Total defence includes passive resistance by the civilian population.[10] If a small country can rapidly mobilise reasonably well-equipped forces supported by the population, it can continue resistance until the political and economic costs to the aggressor exceed strategic benefits.[11]

The Baltic States have collaborated in their defence affairs since they regained independence. The first major project was the Baltic Peacekeeping Battalion or BaltBat, first proposed in a meeting of Baltic Defence Chiefs in 1993. From the start the Battalion had a political aim and a military purpose. Politically, the Baltic States could claim a visible place in international peacekeeping and participate, as sovereign states, in international security matters, that is, engage in military diplomacy. The initial purpose of peacekeeping for the UN has receded into the background while deployment under NATO has moved to the fore. Latvian units have participated in IFOR, SFOR, and KFOR, and in 2003 an infantry company was deployed to Iraq. The Battalion was the first link in a mesh of regional security arrangements. In 1995 the Baltic Defence Ministers signed an agreement identifying the specific areas of cooperation. This led to a Baltic Naval squadron, BaltRon; a Baltic Air Surveillance Network, or BaltNet; and a Baltic Defence College. NATO membership requires interoperable staff procedures, communications systems, similar tactics, leadership principles, and a shared military ethos. English language training has a high priority. Preparation for NATO membership has thus assisted Baltic defence cooperation. All three countries are developing the same command, control, and information systems, logistics, resource management, and training concepts based on NATO experience. Thus external and internal requirements have shaped similar Baltic defence establishments.[12]

In principle, the basic elements of a small conscript army backed by a large reserve force, with strong Baltic defence cooperation and defence development aimed at NATO integration and participation in international peacekeeping missions is a proper response to Latvia's security environment. In practice, an integrated approach to meeting these related, nonetheless challenging requirements was, until the end of the 1990s, difficult to attain. Considering international assistance and requirements, the Latvian armed forces have gained from the process, through experience in Partnership for Peace (PfP) exercises, engagement in the MAP, and participation in multinational peacekeeping operations. However, giving priority to international objectives can place national requirements in a subordinate position to a NATO agenda. General Wesley Clark, visiting the Baltic States in July 1999, favourably rated the progress achieved, commended Baltic defence cooperation, and acknowledged Baltic contributions to IFOR, SFOR, and KFOR. But Clark also advised the defence ministries to weigh their priorities carefully and suggested an emphasis on national military reform and training.[13] Similarly, IDAB recognised that the Baltic governments were enthusiastic members in the PfP and had contributed to international peacekeeping, thus demonstrating their readiness to provide security to others as well as to request it for themselves. 'But we sound a note of caution, as we had done before', noted the Board, 'about the danger of allowing the benefits to be gained from international cooperation to consume a disproportionate amount of the limited defence budget, to the detriment of internal development.'[14] Christopher Donnelly has pointed out that: 'The demands of meeting the requirement to provide competent forces to participate in NATO-led operations ... push a nation down the route of developing forces which are NATO-compatible. But these are so expensive that in order to afford them the country may have to switch scarce resources from a force structure for national defence.'[15] There are two sides to the Western coin.

There is also the question of professionalism. In contemporary writings, the term has different connotations. More narrowly, it means the ability of service personnel, particularly the officer corps, to perform their duties well. More broadly, it refers to professional or volunteer armies versus conscript armies. Latvia has a conscript army. Therefore, professionalism will be dealt with in the narrower definition, the ability of the armed forces to do what is asked of them.

Latvian armed forces are small, with regular and reserve components. Politically, officers have to be conscious that they belong to a democratic society. All of them need a combination of military and civilian skills, including knowledge of the democratic controls placed over the military. Militarily, they conduct operations as platoon, company, and battalion commanders. This places emphasis on unit tactics and communications. At all levels of command, this requires the delegation of command responsibilities, initiative, individual leadership abilities, and good relations between officers and enlisted men—characteristics that were not features of the Soviet model of military professionalism. Company and battalion commanders should have the ability to work with their Baltic counterparts and be good administrators of defence resources in line with relevant legal provisions and procedures. Senior officers serving at higher headquarters, senior staff positions, or at international institutions, are expected to be 'military commander[s], military diplomat[s] (or) ... military policy maker[s]'.[16]

In Latvia's circumstances, military education has had a leading role. Latvia did not have to restructure Soviet military schools since they were removed with the Soviet withdrawal. The National Defence Academy, established in 1992, has undergone several reorganisations. The latest review, completed in 2001, changed the entry requirements, length of studies, and curriculum. The Academy accepts university graduates who have to pass exacting examinations. The cadets receive a short, intensive one-year basic military education, with subsequent assignments at other schools as a part of an officer's career path. The objective is to develop military professionalism in relation to technical military skills combined with high moral standards, intellectual qualities, the ability to lead by example, and good communications with soldiers in the unit. Military discipline and unit cohesiveness are to come from a respectful treatment of soldiers and recognition of individual initiative.[17]

Field grade officers receive education at the Baltic Defence College in Estonia, which was established in 1999. Instruction covers strategy, total and territorial defence, staff duties, and logistics. The students learn operational art, the development and application of military technology, national defence planning, and how planning is related to a nation's economic and social resources. Instruction takes into account the geographic and political conditions, defence concepts, and different administrative and legal systems of each country. Instruction is based on *Auftragstaktik*—mission-oriented command in the defence environment of the Baltic States. The graduates are trained to serve as chiefs of staff at infantry brigade level and at defence regions, for planning positions in defence ministries, General Staff positions, and international duties.[18]

An objective method of promoting officers is clearly a central requirement of professionalism. Donnelly notes, 'as far as defence reform itself is concerned, the most important feature of a personnel management system is that it should deliver the right sort of an officer that the new force structure requires'.[19] Latvia's Defence Ministry has introduced career management for the professional soldiers, with explicit and open procedures, based on established standards of qualifications and achievement. An impartial board with military and civilian representatives evaluates qualifications and makes recommendations.[20] A soldier's career progresses with assignments to command and staff duty and periods of training at the Baltic Defence College or Western military schools. Latvia opened an Academy for Non-Commissioned Officers in 2001.

The Latvian government is making a deliberate effort to accelerate a generational change in the officer's corps, a change in its military culture, coupled with improvements in pay and living conditions for officers, non-commissioned officers, and enlisted personnel.[21] Corruption or misconduct by civilians or the military has cropped up, but where it has occurred, the Defence Ministry has punished it quickly and effectively.[22]

Defence reform was retarded by a lack of clear political support and less visibly a heritage of the Soviet past that undermined professional development. An additional major problem in this area was lack of funds and in particular severe shortfalls in defence expenditure. At the outset, the government's annual budget was small and social needs claimed a large share of what there was. In the absence of any reliable inflow of government revenue there was no dependable expenditure projection and funds voted by the *Saeima* rarely matched that which was needed by the Defence Ministry. While Latvia's defence planners could envisage how the armed forces should be developed, in reality budgets were barely sufficient to keep the defence establishment on an even keel. 'Whilst accepting that there are always competing demands for scarce resources and that the decisions on how these resources are allocated is a political one, to be decided on the cabinet level', wrote IDAB, 'we nevertheless judge that the low proportion of GNP allocated to defence in all the Baltic States has been historically such as to ... frustrate internal military development'.[23] A chronic uncertainty about how much money would be provided for defence made long-term planning a futile exercise. Military and civilians in the defence community developed new skills and marked time, waiting for the dawdling, politician-led supply train to catch up.

However, the late 1990s brought a more serious approach to defence reform. As Latvia's economy began to improve, the *Saeima* was able to vote defence budgets capable of delivering a reform agenda. Moreover, the Latvian government's commitment to NATO membership and encouragement at the Alliance's 1999 Washington summit provided an influential external incentive to turn plans for more professional armed forces into reality. NATO gave a clear message to the government when in an assessment of Latvia it stated that 'the speed at which Latvia can build its armed forces depends in large part upon the level of spending it provides for support as well as the economic rate of growth'.[24] The resolve of the political establishment to commit more fully to defence reform was further bolstered by changing public attitudes to defence. Opinion polls indicated growing support for the armed forces and defence appropriations and an effective public relations programme by the Defence Ministry also helped to shape public views in this area. Indeed, Latvian defence expenditure increased, from 0.69% of GDP in 1997 to 1.3% of GDP in 2001 to 2% in 2003.[25]

Some have suggested—somewhat paradoxically—that Latvia's armed forces would encounter an unusual funding problem because there could be too much money for the armed forces, a glut following very lean years.[26] This view argues that the Defence Ministry might not be able to effectively utilise its budget allocation and that money would be wasted or frittered away. Eventually this may result in public opinion shifting negatively on the matter of defence funding. However, according to the American Kievenaar Study, 'The speed at which Latvia can build its armed forces depends in large part upon the level of spending it provides for support as well as the economic rate of growth.' The study also observed that national defence plans were as basically sound and could be implemented over the next 10 years, provided that funding was made available. Latvia's Defence Ministry has developed and implemented a Planning, Programming and Budgeting System (PPBS), with one-year, four-year, and twelve-year defence planning, and the possibility of a surplus of funds being misspent is not an immediate problem.[27]

International and Transnational Influence

International organisations and various Western states have had a great influence over Latvia's defence development. Given their historical experience and geostrategic situation, all three Baltic States have been anxious to join the Alliance. Consequently, it has had more influence in Latvia than in those countries, like the Czech Republic, where NATO membership was less of a concern. But there is more to the story than requirements, advice, and assistance coming from Brussels on the one side, followed by gratitude, acceptance, and action in Riga on the other. Latvia's defence development began in 1991, three years before the enlargement of NATO and the EU appeared on the agenda. Some of the foreign states that provided aid, or pressure, were NATO members; others, like Sweden and Finland, were not. Moreover, NATO requirements have changed and expanded considerably over time.

Once Latvia decided on NATO membership, it had to conform to layer after layer of requirements and demands, starting with the Washington Treaty of 1949 to the Study on NATO Enlargement of 1995, the MAP and the NATO Strategic Concept of 1999, and others in-between, like the PfP. A country that hopes to join NATO has to observe definite rules of behaviour in its domestic policy and in the relations with its neighbours. It has to have demonstrable civilian control over the military and provide economic wherewithal to reform and sustain its armed forces and there is guidance about defence preparations for membership. States which have ethnic disputes and conflicting territorial claims have to settle their disagreements peacefully. They have to participate in Partnership for Peace exercises, contribute to regional security, and engage in international peacekeeping missions. Fulfilling these conditions did not necessarily guarantee membership. Failing to live up to some could have seen Latvia removed from the list of applicants. As various NATO programmes, requirements, and actions followed each other, Latvia strove to demonstrate its willingness and ability to engage in all programmes and all actions, including IFOR, SFOR, KFOR, and finally sending an infantry company to Iraq in August 2003.

From 1991 onward, external support has been provided using a variety of ways and means. Initially, Western states were circumspect, providing some supplies and some equipment but not weapons or ammunition, because Moscow might object. Activation of the BALTBAT in 1995 brought more support, from the Nordic states in particular. Expert advice was as important as shipments of supplies. The first expert group was an International Defence Advisory Board (IDAB) chaired by General Sir Garry Johnson and composed of retired or senior high level military officers or civil servants from Nordic and some NATO states. IDAB was established in 1995 at the request of the Baltic Defence Ministers. It worked with presidential offices, foreign and defence ministers, chiefs of defence, parliamentarians, senior officials and military officers. There were also guidance reports for each country's defence development, made by an expert team headed by the US Major General Kievenaar. The Danish Ministry of Defence and Chief of Defence Headquarters worked out 'Defence Concepts and Plans' which were presented to the Baltic Security Assistance Group of states (BALTSEA). The main achievements in this period have been the development of national security concepts in all three countries; a reorganisation of Latvia's Defence Ministry and the General Staff; and IDAB representations to NATO and SHAPE to identify what kind of assistance would be of strategic value. Western officers have served as advisers, or directly in the Latvian armed forces. The Baltic Defence College was developed with assistance from the Scandinavian countries. Latvian officers have attended military schools in the United States, Finland, Germany, Sweden, and other countries.[28]

The latest, largest, and most demanding NATO program is the MAP. This requires that a candidate for membership has to provide sufficient funds to reform and sustain its armed forces, engage in PfP, contribute to regional security, and participate in international peacekeeping missions. NATO has indicated that candidate countries should provide 2% of their GDP for defence and many western documents on NATO expansion give the 'two per cent attainment' a prominent place.[29] The MAP is a highly prescriptive document. Each of the aspiring states has to submit annual national plans that cover the full range of their activities in preparation for NATO membership. These plans include not only military measures designed to create force improvements, so as to enable them to carry out the Strategic Concept, but also details of defence resource management and of economic policy. Fulfilling the conditions does not guarantee membership but failing to deliver objectives provides a legitimate reason to delay membership.

The MAP certainly improves defence planning. NATO provides defence planning tools, evaluates a country's progress, and provides technical and political guidance. Latvia submits annual national defence plans to NATO on force improvement, budgeting, and other matters. The Defence Ministry has implemented resource management based on integrated short-, medium-, and long-term plans, adjusted to government budget cycles and the availability of appropriations.[30] In domestic politics, the MAP has certainly proved a useful argument to Latvia's Defence Ministry to secure increased defence expenditure. The Defence Minister's Report to the Parliament of 2000 specifically refers to the need to increase defence expenditure and pledged this by 2002, the year when the Alliance considered candidates for the next enlargement.[31]

There have been disadvantages as well as advantages brought by this swelling wave of outside aid. For example, the International Defence Advisory Board issued its Final Report in 1999. It said that a negative aspect of international assistance was the confusion brought about 'by the plethora of [Western] advice and assistance, often uncoordinated and short-term in nature, offered by supporting nations and organisations'.[32] Similarly, Brigadier General Michael Clemmesen, the Commandant of the Baltic Defence College and the former Danish Defence Attaché to the Baltic States, writes of advisers and support project officers, who are 'unfortunately only too likely to be without prior knowledge or understanding of (Baltic) defence problems', experts who 'only know their own system that mirrors the development of their own forces and the politico-economic and geo-strategic requirements of their own state during the recent years', and simply leave Baltic country with two choices: either copy the supporting state's system fully or lose the opportunity for support.[33] The advisers and support project officers, notes Clemmesen, often do not understand the defence problems of a small, poor, frontline state that is recovering after 50 years of totalitarian, militaristic, corrupt misgovernment.

The Military and Society

Armed forces have histories that give society an image of their nature and purpose. The Latvian view of the nature of armed forces contained different, even conflicting, images. The more recent one was of the Soviet Army. Ethnic Latvians, overwhelmingly—and here one can safely generalise—despised the Soviet Army. Militarily, it was seen as an occupying Army, *okupanti*, which had seized a small independent country in 1940. The mass deportations and executions of the Stalin years were possible only because the Soviet Army had occupied Latvia. Its many large military bases ravaged and despoiled Latvia's natural environment. There was not a single spot in Latvia's past or present to which one could point and say: 'There the Soviet Army did good.'

Another image came from the Latvian past. During the War of Independence Latvia's young national army had fought on two fronts, against the Red Army and German forces, and defeated both. Moreover, Latvian regiments and divisions had served in the Imperial Russian Army during the First World War and in the German Army in the Second World War. They had the reputation of elite combat units. The Latvian historical memory of their armed forces—perhaps oversimplified but not without justification—is one of discipline and valour, recognised by friends and foes.

In the years immediately after independence, there was some public debate about whether Latvia could develop its armed forces so as to safeguard the country's sovereignty. One argument against having any armed forces at all was that Russia was the sole possible aggressor and that a small Latvian armed force could only fight against Russian forces for a short time. The answer to this notion was that the Baltic States had not resisted militarily in 1940 to a Soviet ultimatum, had been militarily occupied, and suffered huge losses through executions and mass deportations. This would never again happen. In the contemporary security environment, Baltic resistance would generate an international reaction. The community of Western states would turn against Russia, and any regime imposed by Moscow would not last long. Moreover, Latvia needed national armed forces to establish law and order in the country, to cope with crisis and emergence situations, and to provide a symbol of sovereignty and a source of national pride.

The Soviet Armed forces left Latvia, but they also left behind an invisible heritage. At the outset, what little knowledge Latvia's civilians had of defence planning was coloured by the Soviet experience. Soviet rule had left a culture that demanded conformity not initiative; control not delegation; compartmentalisation not cooperation; and secrecy not transparency. Although the Baltic political systems had changed, the supporting civil servants were slow to depart from their old ways. Bureaucratic processes did not exist or functioned inadequately and there was a lack of national governmental capacity, of people with overall competence for defence policy formulation and planning. Ministries, parliamentary committees, and presidential advisers often lacked expertise. An aversion to information sharing and a culture of secrecy significantly affected the ability of other governmental agencies to establish relationships with each other and with the media and society at large. What countries like Latvia really needed was a long period of readjustment; time to think out their new national security situation, time and money to plan at a measured pace, time to work out new training systems and procurement policies. But in the real world, everything has had to be done at once, with no clear vision of the future, and with strictly limited funds.

After a decade of development, Latvia's defence community sorted out this confusion, organised coherent armed forces, determined their priorities, put together viable defence plans, and developed a mission statement for them. The sense of history has not been discarded. The armed forces again wear the uniform of the old army, at parades and other public occasions. The gap between the unpleasant Soviet nature of the armed forces in 1991 and armed forces of an independent nation has been largely removed. However, new criteria have emerged with which society judges the armed forces.

Latvia's society is not ethnically homogenous. When we consider the relationship between the armed forces and society, it has to be kept in mind that some 60% of the country's inhabitants are Latvians, 28% are Russians, and 12% belong to other nationalities. In Western studies, those who are not Latvians are usually designated as Russians, but this is not accurate. There are Belarus, Ukrainians, Poles, Lithuanians and Jews who do not necessarily have the same views as the Russians. To divide the entire population into 'Latvians' and 'Russians' as sometimes is done in Western publications is misleading. Nor are the ethnic Russians a grey mass. Some have come to Latvia recently, others have roots reaching back for centuries; there are Russian workers, teachers, professionals, and the intelligentsia; each group has closer or looser ties to society and different views toward Latvia as a state.

Views amongst the minorities sometimes differed from Latvian ones. They had no historical image of a former Latvian Army or, for that matter, of a Latvia that once had been an independent state. They did not view the Soviet Army with intense dislike and contempt. Nor were their views toward Soviet armed forces uniformly positive. The former Soviet military (a handful of them remained in Latvia) longed for past prestige and privileges. As to the others, their attitudes were similar to those in the rest of the Soviet Union, which, after the war in Afghanistan, tended to be critical.

There is no exact way of measuring the responses of a society which is still in the process of consolidation from disparate ethnic groups, as is the case in Latvia. Nonetheless, we have to ascertain as far as possible the society's opinions on the validity of the various missions that the armed forces intend to carry out and their capability of doing that. They can be categorised as functions of national security, nation building, regime defence, domestic assistance, and military diplomacy.[34] Latvia's government has developed its defence policy, starting with a national security concept and followed by other documents and statements, ranging from defence policy to force development plans which, taken together, show in clear detail the missions of the armed forces. Governmental policies and the opinions of society are not necessarily identical. Nonetheless, in a democratic society there cannot be a large gap between what the government requests and what society wants to have.

The gap between the unpleasant Soviet nature of the armed forces in 1991 and armed forces of an independent nation has been largely removed. On the other hand, new criteria have emerged with which society assesses the armed forces. Latvian society evaluates the legitimacy of the armed forces largely in terms of its capability of performing its contemporary missions. Over the decade, a similar change has been emerging among the non-Latvians as that among the Latvians. They also tend to judge the armed forces according to their new missions and functions. For example, what are society's views on sending Latvian troops to Bosnia or Kosovo, Afghanistan and Iraq? Over the years, there have been debates in the Parliament, changing party platforms on defence issue, studies on security affairs carried out by Western and Latvian institutions, and public opinion polls. There is not a great deal of evidence on some issues, more on others, but, in all, some generalisations at least can be made. Taken as a whole, they provide, at least, a sufficient picture.

The major issue is national defence. Latvia's defence planning assumes that Russia cannot be disregarded militarily but that it does not pose an immediate threat to the Baltic States. In this case, Latvian and Russian views in society diverge, but the Latvians do not seem to be particularly perturbed with the external danger. They seem to be more preoccupied with new risks, like organised trans-border crime and corruption. In a recent public survey on the benefits of Latvia's membership in NATO, giving the responses of Latvians and non-Latvians alike, less than 9% said that NATO 'will defend Latvia against Russia' whereas 36% responded in more general terms that NATO 'extends security' or 'precludes conflict'. NATO, evidently, is seen as a security organisation but lacks a sharp definition as a military alliance. To Latvians as a whole, Article V would be a good thing, but it is not immediately necessary.[35]

The majority of the public believes that the defence budget should be increased, the majority of the political parties support increased defence expenditures, and, according law, the annual defence budget will be 2% of the GDP by 2003 and thereafter cannot fall below this level. Experts in economic development—both within and outside Latvia—estimate that Latvia's GDP will continue to increase at a rate between 5 and 7%. The law of defence budgets coupled with a sustained strong economic performance would provide sufficient funds. As matters are at present, there is no difficulty in securing public expenditure for the armed forces.[36]

An energetic public relations programme has shaped public attitudes and opinion polls have indicated growing support for the armed forces and defence appropriations. A comprehensive public survey carried out in 2000 showed that the majority of the country's inhabitants, Latvians, Russians, and other ethnic groups, 51% believed that the defence budget needed to be increased, 26% thought that it was adequate, and 9% thought it could be reduced. Of those surveyed among the other nationalities, 43% said that the budget should be increased. The majority of the younger generation of minority groups, that is people under 24, were in favour of increasing the defence budget. Asked which areas of defence needed funds, 43% of the respondents wanted to ensure the quality of military education, 34% the acquisition of weapons and technical equipment, and 20% thought the improvement of living conditions for service personnel should have priority.[37] Latvia's Ministry of Defence, with justifiable pride, underlined the fact that 51%, the majority of the population, believed that the defence budget should be increased. In this sense, Latvia was unique; even in the United States, it was only after the 9/11 terrorist attacks in 2001 that the majority of the population thought that the country's defence budget should be increased. In the sense of public support, Latvia's armed forces had received strong vote of confidence.

The emergence of a broad and inclusive 'society' is becoming apparent in Latvia. Writing in the journal *NATO's Nations*, Major General (ret.) Dietrich Genschel described the situation in 2002 as one where Latvia 'installed a social integration process, adopted appropriate naturalisation legislation and improved respective procedures. As a result naturalisation is increasing, [Russian] loyalty to the Latvian state is improving, eagerness of the younger generation to participate in a modern, free and civil society is pervasive.'[38] There are differences of opinion among Latvians and Russians, but the notable trend is that Russian views are beginning to resemble Latvian ones. If anything, society as a whole seems to be concerned more with the new risks (trans-border crime) than traditional threats (trans-border military attack). Latvia's society evaluates the legitimacy of the armed forces largely in terms of its capability of performing its contemporary missions. Over the decade, a similar change has been emerging among the non-Latvians as the Latvians. They also tend to judge the armed forces according to their new missions and functions.

There are two major issues which indicate society's support for its armed forces. For Latvia, and the Baltic as a whole, joining NATO, a sensitive issue in domestic politics, highlights views among ethnic and political groups. Surveys conducted in 2000 and 2001 showed that 68% of Latvians favoured joining NATO and approximately 44% of Russians. One influential study concluded that Russian support is 'a high figure considering the enormous anti-NATO propaganda in the local Russian press and from Russia itself', which paints NATO in the darkest colours as an aggressive military alliance meant for waging war.[39] Efforts by the Latvian government to rebut the propaganda have been successful. As a result, support amongst Russians is advancing faster than among Latvians'.[40] After Moscow ended opposition to Baltic NATO membership, Russian acceptance of the Alliance continues to increase rapidly. Amongst the younger Russians, in the age group from 18 to 30, it is 62%. After the Prague summit, the latest poll of June 2003 conducted by market and public opinion research company *Latvijas fakti* (*Latvian Facts*) shows that nearly two thirds of Latvia's population support Latvia's membership in NATO.[41]

Conclusion

The period since 1991 has seen remarkable change in Latvia. The armed forces have been built from virtually nothing, a Defence Ministry has been created, a constitutional and legal framework for democratic civil–military relations has been established, and military–society relations—which during the Soviet period were characterised by deep hostility and mistrust—have been transformed. Behind a new defence structure—the units of the Armed Forces, special formations like the BaltBat, new training and education establishments, there is an entirely new approach to defence and security affairs, which, broadly, we can call a 'Western' one. It entails budgeting, programming and planning skills, resource based force development plans, and military strategies and defence policies that are based on national security concepts. It functions because the Soviet heritage, the mentality of bureaucratic conformity, centralised control, delegation, compartmentalisation and secrecy, has been swept away.

Latvia's success in establishing and developing its national defence system to the standards required by NATO membership has occurred for a number of reasons. The apparent disadvantage of having no established defence sector, and no established civilian or military cadres in this area has worked to the country's advantage, by minimising some of the more negative communist legacy issues that have inhibited defence reform processes in other postcommunist states. While Soviet traditions and practices were troublesome during the first years—as some of the problems discussed illustrate—this situation required Latvia to plan its defence reforms in a very deliberate and explicit manner. For both political and geostrategic regions, Latvia identified membership in the EU and NATO as key foreign policy goals early on. This close engagement in the NATO accession process has had two main positive impacts. It gave the Latvian defence reform process a defined—though not always consistent—set of targets to aim for, in relation, for example, to democratic control of the armed forces or specific military capabilities. It also allowed Latvia to receive extensive Western assistance quite early on in its defence reforms. While the impact of this assistance has sometimes been rather mixed, it did help Latvia to make up for its own expertise and experience gaps in the defence sector. Moreover, a considerable proportion of Latvia's new military and civilian defence community has experienced 'the West' directly and personally, at schools abroad, NATO and SHAPE Headquarters, and IFOR, SFOR, and KFOR duty tours. Finally, since independence, the military have played a variety of roles that have both increased their standing in Latvian society and helped create a societal consensus on the general direction of defence reform. In the early 1990s for example, they played an important symbolic nation-building role, while more recently they have been at the vanguard of the country's European integration efforts through their key role in the NATO accession process.

During the past thirteen years Latvia achieved two great aims. Its accession to NATO and to the European Union in 2004 represent the successful culmination of thirteen years of work in its defence sector and civil–military relations. Unlike NATO, the European Union has not been explicitly a part of this study; nonetheless, its increasing attention to security has moved it closer and closer to NATO defence concerns and requirements. Membership in the two great international organisations provides Latvia and all three Baltic States with great external security against an old threat from the East. Moreover, the peoples of the three countries see 2004 as the year when they regained full membership in Europe. It was said by a Latvian observer that 'for us, World War II began in 1939 and finally and completely ended in 2004'. There is much truth in this comment. It could be said that the door was closed to the tragedies and dangers of the twentieth century. However, another door was opened to the risks and threats of the twenty-first century. Because Latvia now is a member of the Euro-Atlantic security community, it has to undertake new obligations and accept future requirements. A Report of the Defence Minister to the *Saeima* in 2001 has a chapter 'Looking to the Future'. It is introduced with a photograph showing the Monument of Liberty at the very centre of Riga, the symbol of independence, built during the 1930s, never destroyed but neglected under Soviet rule, and recently restored to its former grandeur. People are placing flowers at the base of the monument; they are commemorating independence regained. A young sentry stands on guard, wearing the high collared, dark green tunic of the armed forces worn by the Latvian army before 1940. He looks forward with confidence. A little light-haired girl attentively and inquiringly observes the young sentry. We can say that she symbolises society and the future. But what does the future hold?

The future development of the Latvian national defence system is now taking place in the context of full NATO membership. On the one hand, even as its new defence community continues to develop national defence capabilities—a process well under way but not complete—it has to envisage training, deploying, and sustaining forces for participation in multinational missions in the NATO or EU framework to meet the security challenges of the twenty-first century. This could be problematic. In particular, the resources available for the continued development of the armed forces remain limited—despite recent increases in the defence budget. Security policy options will have to be estimated taking into account defence capabilities. However, Latvia has acquired and assimilated the ways and means of accomplishing that, a broad spectrum that starts with national security concepts, debated in the *Saeima* and known to society, and reaches down to useful briefing charts prepared by the military and the civilians for the Defence Minister. Given the base from which it started—26 sunken Soviet submarines and ships leaking acid, oil, and phosphorous—Latvian progress in defence reform has been remarkable.

Notes

[1] NATO Parliamentary Assembly, Committee Reports, *European Security: The Baltic Contribution* (November 1998), p. 12.

[2] The International Defence Advisory Board for the Baltic States (IDAB), *Final Report*, (February 1999). The Board was established in 1995 at the request of the Baltic Defence Ministers to advise and assist with Baltic defence and security reform.

[3] Andrew Cottey, Timothy Edmunds & Anthony Forster, 'Introduction: The Challenge Democratic Control of Armed Forces in Postcommunist Europe', in Andrew Cottey, Timothy Edmunds & Anthony Forster (eds), *Democratic Control of the Military in Postcommunist Europe: Guarding the Guards* (Houndmills: Palgrave, 2002), p. 4.

[4] These are: *The National Security Law* (2001); *The Mandatory Military Service Law* (1997); *The Law on the National Guard of the Republic of Latvia* (1993); *The Law on Participation of the National Armed Forces of Latvia in International Operations* (1995); *The Law on the National*

Armed Forces (1999); *The State Defence Financing Law* (2001); *The Law on Status of Foreign Armed Forces in the Republic of Latvia* (1997); and *The Alternative Service Law* (2002).

[5] See NATO Parliamentary Assembly, *The Baltic Contribution*, and the findings of an expert group, headed by Major General H. A. Kievenaar, US Army, in the Baltic States, *Latvian Assessment Document (NATO Unclassified)* or the 'Kievenaar Study', 1997. Available at http://www.naa.be (accessed).

[6] International Defence Advisory Boad, *Final Report*, p. 6.

[7] Republic of Latvia, Minister Cabinet, *Security Concept of the Republic of Latvia* (2002).

[8] IDAB, *Final Report*, p. 4.

[9] Republic of Latvia, Minister Cabinet, *The National Defence Concept of the Republic of Latvia* (6 June 1997). This document presents in more detail the defence aspects of security outlined in the security concept.

[10] Republic of Latvia, Defence Ministry, *Basic Principles of the Defence System of Latvia. Total Defence and Territorial Defence* (1999); and Michael H. Clemmesen, 'Territorial Defence in the Baltic Defence College,' *Baltic Defence Review*, 3 (2000), pp. 83–86.

[11] See K. G. H. Hillingso, 'Defensibility', *Baltic Defence Review*, 1 (1999), pp. 1–4; and Robert Dalsjo, 'Baltic Self-Defence Capabilities—Achievable and Necessary, or Futile symbolism?', *Baltic Defence Review*, 1 (1999), pp. 11–15.

[12] Eitvydas Bajarunas, 'Baltic Security Co-operation: A Way Ahead', *Baltic Defence Review*, 3 (2000), pp. 43–62.

[13] *Central European Review*, 9 August (1999), p. 23.

[14] IDAB, *Final Report*, p. 6.

[15] Christopher Donnelly, 'Reshaping Armed Forces for the 21st Century', *NATO Think Piece*, 10 August (2001), p.14. Available at http://www.nato.int/docu/articles/2000/a000913a.htm (accessed). See also Michael H. Clemmesen, 'Before Implementation of the Membership Action Plan: Baltic States? Defence Development Until the Present', *Baltic Defence Review*, 2 (1999), pp. 35–42.

[16] Baltic Defence College, *Information*, Department of Strategy and Political Studies, (1999).

[17] Ilmars Viksne, 'Formation of the NDA', *Baltic Defence Review*, 3 (2000), pp. 17–29; and Republic of Latvia, Defence Ministry, *Military Education System at the National Defence Academy*, November (2000).

[18] See M. Clemmesen 'The Development of Regular Army Officers—An Essay', *Baltic Defence Review*, 3 (2000), pp. 7–16.

[19] Donnelly, *Reshaping European Armed Forces*.

[20] Republic of Latvia, Defence Ministry, *Aizsardzibas ministra zinojums Saeimai par valsts aizsardzibas politiku un Nacionalo brunoto speku attistibu 2000. gada*, pp. 44–48.

[21] Republic of Latvia, Defence Ministry, *Aizsardzibas ministra zinojums Saeimai*, pp. 84–86.

[22] Dietrich Genschel, *High Level Leadership within the Defence Establishments of Democratic Societies*, Lecture, Tartu University, Estonia, February (2000). General Genschel is Germany's representative on the International Defence Advisory Board.

[23] IDAB, *Final Report*, p. 5.

[24] *Latvian Assessment Document*, p. 93.

[25] *Defence Budget of Latvia in per cent of GDP*, Latvian Ministry of Defence Website. Available at http://www.mod.lv/english/03budzets/01_01.gif (accessed).

[26] Donnelly, *Reshaping European Armed Forces*.

[27] 'Kievenaar Study', 1997.

[28] See Robertas Sapronas, 'BALTBAT and the Development of Baltic Defence Forces', *Baltic Defence Review*, 2 (1999), pp. 55–70; and T.D. Moeller, BALTBAT—Lessons Learned and the Way Ahead', *Baltic Defence Review*, 3 (2000), pp. 38–42.

[29] See *Debate on NATO Enlargement, Hearings Before the Committee on Foreign Relations, United States Senate* (Washington: US Government Printing Office), pp. 439–454, where the 2% achievement appears as an important consideration.

[30] Republic of Latvia, Defence Ministry, *Latvia's Membership Action Plan 2000*; and *Membership Action Plan. Latvia's Annual National Program 2001*.
[31] Republic of Latvia, Defence Ministry, *Aizsardzibas ministra zinojums Saeimai*, pp. 25–30; and Republic of Latvia, Defence Ministry, *Trends of Defence Budget for 1999–2003* (1999).
[32] IDAB, *Final Report*, p. 6.
[33] M. H. Clemmesen, 'Supporting States Advice and Defence Development', *Baltic Defence Review*, 4 (2000), pp. 9–10.
[34] For more on this typology see, Timothy Edmunds, Anthony Forster & Andrew Cottey, 'Armed Forces and Society: A Framework for Analysis', in Anthony Forster, Timothy Edmunds & Andrew Cottey (eds), *Soldiers and Societies in Postcommunist Europe: Legitimacy and Change* (Houndmills: Palgrave-Macmillan, 2003), pp. 9–14.
[35] *Public Opinion Survey 2000. Attitude toward military and defence issues*. Available at http://www.mod.lv/english/02.darbs/04.sab.php (accessed).
[36] Ibid.
[37] Ibid.
[38] *NATO Nations*, Special Issue. Latvia (2001), p. 3. Major General Genschel, retired from the *Bundeswehr*, was a member of the International Defence Advisory Board.
[39] *Assessing the Applicants for NATO Membership: Final Report of a Study on the Membership Action Plan (MAP) States and 'Second Wave' NATO Enlargement* (Groningen, Netherlands: Centre for European Security Studies, July 2001).
[40] Ibid.
[41] Latvia, Ministry of Foreign Affairs. Available at http://www.mfa.lv (accessed).

Civil–military Relations in Croatia: Politicisation and Politics of Reform

ALEX J. BELLAMY* & TIMOTHY EDMUNDS**
*School of Political Science & International Studies, University of Queensland, Queensland, Australia **Department of Politics, University of Bristol, Bristol, UK

Civil–military relations in Croatia are receiving an unprecedented level of international interest. According to Misha Glenny, the question of defence reform in Croatia is of crucial importance for the rest of the Balkans. Because it has a more advanced economy than all of its southern neighbours, Glenny argues, if Croatia cannot successfully reform to suit the demands of NATO and the European Union (EU) there is little hope for the rest of the region. Moreover, he suggests, if Croatia's reform process fails, Western leaders are likely to become more sceptical about the chances of success in states like Serbia and Montenegro, Albania, and Bosnia and Hercegovina (BiH).[1] Domestically, defence reform remains a key element of Croatia's bid for NATO membership. Along with membership of the EU, this is the centrepiece of the current government's wider foreign policy agenda. This article aims to contribute to this emerging debate on Croatian civil–military relations by tracing their evolution since independence and outlining the prospects and potential pitfalls contained in the current reform agenda.

Civil–military Relations since Independence

Civil–military relations in Croatia have passed through three phases: nation building between 1991 and 1995; regime defence between 1993 and 2000; and reform between 2000 and the present.[2]

Between the dissolution of Yugoslavia in 1991 and the end of the Bosnian war in 1995, the role of the Croatian armed forces can be best described as *nation building*. Between 1991 and 1993, this role focused on the physical defence of national institutions and territory. The military was at the vanguard of developing the Croatian state in response to an immediate threat to state survival. The Croatian armed forces were constructed from scratch, in time of war, to fulfil an immediate functional goal.[3] In 1991, in response to Croatia's declared intention to seek independence from socialist Yugoslavia, Serb militia and the JNA launched a series of attacks throughout Croatia, culminating most famously with the sieges of Vukovar and Dubrovnik. Serb forces seized around one-third of Croatia's territory and at times the front line came as close as thirty kilometres from Zagreb. In response, the new Croatian state had to fashion its armed forces on the basis of the disparate resources that it had at its disposal. These included local police forces, former members of the Yugoslav National Army (JNA) and civilian volunteers.[4]

It was not until September 2001, when the war was already several months old, that a functioning General Staff and military hierarchy were created. During this formative period, processes of state building and military building went hand in hand and civilian and military authority was often indistinguishable. Even once a structure resembling a regular army was created, significant paramilitary organisations remained outside state control and the police force continued to undertake military roles.[5] Nevertheless, as one of the earliest forms of state institution, the armed forces played a crucial role in both protecting the state and society from attack and building the physical capacity of the state.

Towards the end of this first period, the emphasis of the nation building function began to move away from institution building towards supporting the nationalist policies of President Franjo Tuctman. Throughout the 1990s, the Tuctman government attempted to enforce its own particularist vision of Croatian national identity.[6] According to the government, this identity was Catholic, historicist, conservative, and would extend into western Hercegovina. Accordingly, in 1993 the Croatian army was used to assist Bosnian Croats in their fratricidal war against the Bosnian Muslims. The use of the Croatian military in this way prompted controversy in Croatia proper and began damaged the military's standing in society.

The second phase of civil–military relations in Croatia can be traced from 1993 to the beginning of 2000. During this period the military primarily fulfilled the role of *regime defence*. Indeed, once the perception of national emergency had receded, it became evident that the Croatian military had been heavily politicised and was being used in a variety of ways to support the ruling party.[7] The President created his own praetorian guard; intelligence services were often used to monitor the activities of political opponents and undermine the work of reformists within the ruling party; and the Ministry of Defence and armed forces used to deliver patronage to the government's political allies. Within the defence sector, budgeting was inaccurate, and corruption, profligate spending and wastage were endemic.[8]

In combination with the legacy of the war and the authoritarian governance of the Tuctman regime, this tendency towards illiberal and undemocratic patterns of civil–military relations contributed significantly to Croatia's increased international isolation. Whilst states such as Slovenia, Macedonia, Albania, Uzbekistan, Kazakhstan and others were admitted to Partnership for Peace, Croatia remained excluded. The EU retained its arms embargo on the county—which severely restricted defence cooperation (and prevented it altogether until 1998)—until November 2000.[9] Although the US did engage in defence diplomacy relations with Croatia, and Croatian officials and defence experts joined international epistemic communities, these remained 'arms length' relationships. For much of this period, defence assistance in areas such as education, planning and budgeting were only offered through ostensibly independent organisations such as Military Professional Resources Institute (MPRI).[10]

During this period, the armed forces were deeply penetrated by the ruling Croatian Democratic Union (HDZ) party and defence spending remained disproportionately high.[11] This funding did not, however, go towards the creation of professional, capable or democratically controlled armed forces. Indeed, prior to 1998 Ministry of Defence spending was not even audited by the State Auditing Office and nor did it provide yearly reports to parliament.[12] The Ministry of Defence, in common with the rest of the state bureaucracy, was also hugely overstaffed.[13] This endemic politicisation soured relations between the military and wider society in Croatia. Indeed, relations got so bad that there were genuine fears that if Tuctman died and the HDZ lost the 2000 elections (as subsequently happened) there would be a military coup led by hardliners. Although these elections were verified as being 'free and fair' by the Organisation for Security and Cooperation in Europe (OSCE), observers noted that military personnel were put under considerable pressure to cast their vote for the HDZ and that the secret ballot was not always practised in military barracks.[14]

The third, and current, phase runs from the beginning of 2000 to the present and can be labelled a period of *reform*. This period began with the defeat of the Tuctman-era HDZ government by a six-party coalition led by Ivica Račan in 2000. It also includes the period from November 2003 when a reformed HDZ party led by Ivo Sanader swept the Račan government from power.[15] During this period, and although change has been slower and less coherent than had been hoped for, there have been dramatic changes in patterns of civil–military relations. In particular, the Croatian armed forces have taken on a range of new roles, whilst both the Račan and Sanader governments and military have worked to establish a properly democratic control over the defence sector. All these changes have taken place in the context of expanding cooperation with NATO and active engagement in its Partnership for Peace (PfP) programme as well as associated activities.[16]

The reform process has also included a rapid reconsideration of the relationship between the military, the parliament, the presidency and the government. Although bureaucratic reforms have created confusion about chains of command and responsibility (see below, p. 76–77) and have delayed defence reform more generally, the military has been wrenched away from both the ruling party and particular government institutions. The development of a National Security Strategy (2002), Defence Strategy (2002) and National Military Strategy (2003) as well as four other defence-related laws have all contributed towards the democratisation of civil–military relations.[17] These reforms have also had a palpable effect on Croatia's international relations. At the governmental level, Croatia has now joined a whole host of organisations, acceded to a NATO Membership Action Plan (MAP) in May 2002, and was invited to start negotiations for EU membership in April 2004. These changes have also spawned a wealth of transnational relationships between the defence community in Croatia and similar communities overseas. Defence intellectuals and defence ministry officials regularly attend conferences and courses overseas. Elements of the Croatian military collaborate internationally in a variety of ways that include joint exercises in the framework of PfP and bilateral activities with countries such as the US, Germany, Hungary and Italy. The Armed Forces of the Republic of Croatia (OSRH) also send officers on overseas training and education programmes.[18]

Civil–military relations in Croatia have therefore evolved considerably in their first decade or so, and remain in a state of change. In many areas the military is at the vanguard of transition, for example by supporting Croatia's aim of European integration through its key role in the NATO accession process. However, the continuing legacies of authoritarianism and conflict, as well as the political fragility of the Račan government whilst in office, have created a number of continuing problems for the reform agenda, causing it to almost stall completely at times. The following section identifies five of the key impediments to reform in Croatia.

Impediments to Reform in Croatia

The defence reform agenda is a global phenomenon that creates expectations about what a state's armed forces ought to do and the *processes* of defence decision-making.[19] Since 2000, the Croatia's attempts to translate and implement this global agenda have highlighted a number of key constraints on reform. These suggest a need to rethink aspects of the reform agenda and develop a more contextually sensitive approach that takes greater account of local interests and values. Five types of impediment, in particular, can be identified: political, institutional, societal, economic and international.

Political Impediments

The political dimension of the reform agenda focuses on the creation of democratic oversight of the armed forces.[20] Such oversight requires civilian authorities to control the security sector, demands that the political system be democratic, that elected bodies exercise oversight over the security sector, and that the process be accountable to civilian authorities, wider society and the international community. Political impediments to reform can inhibit the creation and maintenance of civilian and democratic control over the security sector and limit the capacity of institutions to fulfil their role effectively and efficiently.

In the Croatian case, there have been five primary political impediments to defence reform. The first is an absence of clear separation of powers between civilian and military authority. Although Zvonimir Mahečić was correct to argue in 2002 that 'there is no serious highly positioned professional in the Croatian army who would question the right of politicians to play the key role' in defence management, he went on to argue that 'the biggest threat to democratic civil–military relations in the last three years has come from ill-prepared politicians.[21] Thus, although there is rhetorical support for civilian control, in practice there remains considerable suspicion within the military about the role of civilians and there is a dearth of civilian expertise about military matters.

The second key political impediment was the divided nature of the Račan coalition government that ruled between 2000 and 2003. The government consisted of five (originally six) parties, including socialists, liberals, conservatives and regionalists and a President and Prime Minister from different parties.[22] During this period, there was no coherent government agenda and each new piece of legislation was the product of months of debate—often reflecting the lowest common denominator that all members of the coalition could subscribe to. In this climate, creating and implementing a strategic vision on defence reform proved extremely difficult. For example, it took over two years for the government to agree and introduce key defence legislation such as the National Security Strategy and the Defence Law.[23] While the election of the new HDZ government in 2003 on the basis of a far stronger coalition may now have addressed this particular problem, it is clear that the weakness of the previous government has significantly delayed Croatia's defence reforms.

The third key problem is the legacy of politicisation. While political activity within the Ministry of Defence was officially discontinued in 2000, supporters of the HDZ and Tuctman-era appointees remained preponderant within the state and military bureaucracy. Faced with this challenge, the Račan government could not adequately resolve the dilemma of how to 'de-politicise' the defence sector (remove political appointees) without 're-politicising' it (appointing a new generation of government appointees) and itself was accused of rampant politicisation by its opponents.[24] In particular, the diverse nature of the Račan coalition meant that key jobs tended to be divided up on between different parties rather than filled on the basis of suitability or professional competence. The new Sanader government has likewise been accused of making a number of politically inspired appointments and dismissals within the Ministry of Defence.[25]

Fourth, Croatia lacks independent civilian defence expertise, particularly amongst its parliamentarians. This means that although parliament has the constitutional right to scrutinise defence policy it is often unable to do so thoroughly because of its lack of technical expertise.[26] Similarly, lack of civilian expertise in the Ministry of Defence has hampered its effectiveness and occasionally led to tensions with military officers in the General Staff. Fifth, related to this, Croatia lacks a developed non-governmental community in defence that has technical expertise and the freedom to question security policy. While the University of Zagreb, several research organisations and a number of extremely capable civilian experts do work on defence and security issues in Croatia, they remain relatively small in number, are often centred around individual personalities and are based overwhelmingly in the capital.[27]

Institutional Impediments

Defence reform is concerned with creating institutions which accept that their role is to carry out the wishes of the legitimate government in an effective and efficient manner.[28] This requires that institutions such as the Ministry of Defence, General Staff, and individual parts of the defence sector have the capacity to fulfil their tasks and remain accountable. Institutional impediments to reform reduce the state's capacity to fulfil its security tasks but also inhibit the ability of civilian authorities to maintain their control over the security sector.[29]

In the Croatian case, the primary institutional impediments to reform are the legacy of politicisation and lack of expertise discussed earlier and confusion and complexity in bureaucratic and political chain of command. In particular, for much of the early reform period (until 2002 at least), confusion and conflict over the relative power and responsibilities of the Presidency, Parliament, Ministry of Defence and General Staff stymied the reform agenda. Under the Tuctman regime, the whole apparatus of government revolved around the President and the elite networks gathered around him. This network of governance and corruption was informal. The parliament held the formal reins of power but was unable to exercise them in practice and even the former Prime Minister had little or no say in the matter. The one area that presidential control was cemented in law was defence. The President, as Commander-in-Chief of the armed forces wielded direct control over all aspects of defence.[30]

The presidential election in 2000 can be characterised as a contest over who would be least like Tuctman and who would relinquish the most of these formal and informal powers. Stipe Mesić won this contest and although he has divested himself of many of his powers and has abolished institutions of presidential control such as the National Security Committee, he has fought to retain his role as Commander-in-Chief.[31] In practice therefore, the General Staff now works to the President and significant tensions exist between it and the Ministry of Defence. The most striking example of this division occurred in 2001 when Defence Minister Jozo Radoš announced that he intended to abolish conscription by the end of the year without any prior consultation with his military colleagues in the General Staff.[32] Similar institutional complexity is evident in relation to senior military appointments. These are made by the President in consultation with the General Staff. Parliament has a right of indirect oversight and the Ministry of Defence has virtually no input. Senior civilian appointments however are the almost sole remit of the ministry.[33]

In the context where the President, Prime Minister and Defence Minister may be from different political parties and ideological backgrounds this institutional complexity risks producing politicised appointments and competition between different bureaucracies. The appointments system provides only one example of how this confused institutional context is inhibiting reform. On issues such as conscription, professionalisation, procurement and the promulgation of a new defence law (eventually passed in 2002) the divisions have been just as pronounced and equally stultifying.[34] Once again, just as with the political context, Croatia's institutional context has made the articulation and implementation of a coherent strategic vision difficult. Instead, defence laws have been based on the lowest common denominator, while planning has been ad hoc and oriented towards crisis management.

Societal Impediments

The defence sector does not fulfil a purely instrumental role in maintaining law and order and protecting the state from external threats. Armed forces also play important social roles such as nation building, providing employment, assisting the construction of infrastructure (such as roads and railways) and providing education.[35] They are not divorced from the societies they operate within. The level of internal legitimacy of the defence sector therefore plays an important role in shaping societal perceptions of defence reform. By challenging its legitimacy, societal impediments can increase the political costs of the reform process, reducing its pace and impact. In the Croatian case, there are three primary societal impediments.

The first is a societal division about the standing of the OSRH. This split emerged as the role of the military subtly shifted during the nation building phase between 1991 and 1995. It was further aggravated during the time of the

Račan coalition by the vocal opposition of the HDZ and other conservative groups to most aspects of the government's defence reform plans. For these groups and their supporters, the defence reform process under the Račan government constituted an attack on the memory of the Homeland War and the very institution of the military.[36] On the issue of cooperation with the International Criminal Tribunal for the former Yugoslavia in the Hague (ICTY), for example, a popular conservative mantra is that because Croatia's war was a defensive one war crimes could not have been committed by Croats. On defence budget restructuring, conservatives argue that cuts only provide evidence of the government's anti-military stance and that attempts to weed out corruption and HDZ patronage merely constitute the repoliticisation of defence. Finally, on the issue of European integration they argue that the government is being at best naïve and at worst treacherous. They insist that reliance on the West for territorial defence jeopardises the very existence of the state. Ironically, since winning the elections in 2003, the new HDZ government under Sanader has taken some impressive steps in actually implementing reforms in all of these areas. Observers in Zagreb suggest that it has been able to do so precisely because its impeccable nationalist credentials have insulated it from criticism from conservative sections of Croatian society.[37]

Other significant elements of Croatian society—particularly liberals, socialists and the urban young—have traditionally regarded the military with deep suspicion and see the defence budget as a potential area for deep cuts to support the extension of social spending or tax cuts.[38] For many, there is also a continuing perception that the military is a haven for right-wing politics, a view which has not been helped by a continuing institutional reluctance to cooperate with the ICTY, the right-wing and overtly pro-HDZ credentials of the many veterans associations, and widespread anti-government campaigns spearheaded by people associated with the military.[39]

The second key societal impediment relates to the legacy of the Homeland War. Precisely because the Croatian state was established in a national war for survival there remains a continued societal sensitivity to external threat. For most Croatians', the primary threat for most of the 1990s emanated from Serbia. Despite the ousting of Milošević in 2000 and an increasingly benign interpretation of the regional security environment by the government and much of Croatian society,[40] the memory of the Homeland War remains strong for many Croatians, particularly conservatives and war veterans. For them, post-Milošević Serbia is not that much different from the old one in many respects—a perception which was recently reinforced by the strong showing made by nationalist candidate Tomislav Nikolić in Serbia-Montenegro's presidential elections in June 2004.[41] There is therefore a continuing perception amongst some sections of society that the armed forces need to retain a privileged position *vis-à-vis* civilian authorities. This in turn has made it politically expensive to pursue certain reforms that may be regarded as damaging to the military.

Economic Impediments

The 'economic dimension' of defence reform is concerned with the appropriation of resources by the defence sector, mechanisms for revenue collection and the wider economic impact of defence management and defence activity.[42] Defence reform aims to establish transparent defence budgets, an economically sustainable defence system, and the efficient and effective use of resources by the defence sector. Economic impediments are perhaps the key constraining factors in Croatia, and the defence sector has struggled to cope with the wider challenges of post-conflict and post-authoritarian reform in an environment of successive defence budget cuts.

In 2000, the Ministry of Defence and armed forces were identified by the incoming government as an area of profligate spending. Initially, savings were made by the reduction of the number of cars in the Ministry's fleet (there was an estimated surplus of around 2,000) and the removal of privileges such as government serviced credit cards that were used extensively for entertainment, mobile telephones, car accessories and gifts.[43] However, these measures failed to address the underlying structural costs and inefficiencies in the Croatian defence sector and economic constraints continue to hamper the reform process. The defence budget has suffered year-on-year cutbacks both relative to other ministries and in real terms, falling from 6.084 billion Kuna or 4.4% of GDP in 1998 to 3.99 billion Kuna or 2.03% of GDP in 2003.[44]

Perhaps most significantly however, the structure of the Croatian defence budget remains heavily skewed towards personnel and operational costs, leaving little left over for investment in areas such as procurement or education. Indeed, in 2002 Defence Minister Radoš claimed that 67.9% of the defence budget was allocated to personnel costs and 21.4% towards operating costs, leaving only 10.7% for all the defence sector's other expenses.[45] In order to reduce the proportion of the defence budget spent on personnel, both Račan and Sanader governments have embarked on a significant process of downsizing the OSRH. This will be discussed in greater detail below. However, for now it is enough to note that the downsizing programme itself has—in the short term at least—added significantly to existing demands on the state budget as it has also had to fund a generous resettlement package for departing soldiers.

International Impediments

By creating transparent and professional armed forces, defence reform aims to promote international security and through military cooperation, confidence building, and multinational activities such as peacekeeping. However, there are a number of significant international impediments to defence reform. In particular, Croatia is a major recipient of Western defence diplomacy activities, both through NATO and bilaterally. This carries with it a requirement to prioritise areas of comparative advantage or areas that NATO ask for work on. As a result there is a direct correlation between areas singled out by NATO for defence diplomacy—in particular the capacity to contribute units to multinational military operations—and areas where practical reforms and investment in the armed forces have been most apparent.[46] Such an approach risks creating a selective and piecemeal approach to reform that is more geared towards the demands of the international community than to Croatia's own domestic requirements.

Moreover, as a recipient of defence diplomacy, Croatia has been offered a variety of ostensible solutions to its defence reform challenges. However, there has often sometimes been little coordination of the international engagement with the defence reform process in the country. For example, because of the EU arms embargo, which until 1998 included a ban on all forms of military and defence assistance, Croatia's international defence cooperation was initially based upon its relationship with the United States and MPRI.[47] As a result, Croatian views on defence management closely mirrored American thinking. MPRI was contracted to provide military training and education, which included a programme of reshaping the democratic control of armed forces that formed part of the Long Range Management Programme. As a result of this influence, the General Staff and Defence Ministry were organised into huge structures of eight departments each. These were modelled on the Pentagon and bore little relation to Croatia's needs. Indeed, one of the most important steps taken by the Croatian Ministry of Defence in 2003 was an internal reorganisation which rationalised this structure down to a more manageable four departments.[48] Since the EU lifted its embargo, a number of alternative approaches have presented themselves, both increasing the potential relevance of externally funded projects but also raising the spectre of incoherence and duplication as different governments and organisations pursue different agendas.

These impediments to defence reform in Croatia are both material and ideational. They increase the political costs of particular aspects of reform; reduce the capacity of bureaucracies to create and pursue strategic visions; and risk the development of 'patchy professionalisation'. Defence reform in Croatia has therefore taken place against a background which constrains both its speed and scope. Bearing this in mind, the following sections focus on two key areas of the reform process: the institution of democratic control of the armed forces and their professionalistion.

Establishing Democratic Control of the Armed Forces

One of the abiding lessons that Croatia has to teach us about the way that we think about civil–military relations in general is that it is not necessarily *civilian* control of the military that is important, it is the question of how *democratic* that control is that is central. During the 1991–5 war, the armed forces were personally controlled by President Tuctman and his maverick Defence Minister Gojka Šušak. As Ozren Žunec has pointed out, 'civilian control of the armed forces was established through politicization of the armed forces via the penetration model [ie. the penetration of party officials into the defence sector]'.[49] This process of politicisation took two principal forms. First, the political elite infiltrated the military with its own supporters creating a cadre of government supporters at every level of the defence sector. Second, key members of the governmental system claimed democratic legitimacy to impose their own form of personal rule on the armed force, establishing, amongst other things, an elite presidential Praetorian Guard and a complex network of intelligence agencies that carried out the ruling party's bidding.[50]

At the very beginning of the 1990s, the centre exerted very little authority over the armed forces, which had been created out of local militias and other volunteer groups to meet an urgent need. This situation began to change when Tuctman replaced his first Defence Minister, Martin Špegelj, an experienced military commander, with Šušak, a returning émigré and radical nationalist with no military knowledge or experience. Between them, Šušak and Tuctman shaped civil–military relations for the rest of the decade, ensuring that the armed forces were not open to democratic scrutiny or control and turning them into armed wing of the regime.[51]

Although Croatia's new constitution created a semi-presidential system, the President was installed as Commander-in-Chief of the armed forces. Tuctman thus exercised extensive control over the development of the HV. The President also established a Defence and National Security Committee (VONS) whose members he appointed personally, and which allowed him to side-step parliamentary scrutiny of his command of the military while operating under the emergency powers granted to him by the constitution for use in time of war.[52] By subverting mechanisms for the parliamentary scrutiny of the armed forces, Tuctman and Šušak created an environment that allowed them to use the military to pursue the interests of the ruling regime rather than the interests of the state—a fact exposed in 1993 when the HV was used to attack Bosnian Muslims without the government indicating a shift in its Bosnian policy or parliament being consulted. During this first period, then, the HDZ moved to rapidly politicise the armed forces and conflate the separation between regime, state, and military. It is worth quoting Biljana Vankovska-Cvetkovska at length here:

> The leading positions in the military were filled with political activists of the HDZ, with almost no military education of professional experience. Thus, direct political influence and control over the military were established from the very beginning. The control has been strengthened by the fact that the position of the Commander-in-Chief has been held by the leader of the ruling party. Many Croat officers confirm (unofficially,

of course) that the majority of the members of the military staff are HDZ members. Many of them were forced to join the party.[53]

In addition, a military counter intelligence organisation (SIS) was established which allowed the ruling party to penetrate and monitor the military at all levels.[54]

During the second phase of Croatian civil–military relations, the politicisation of the armed forces became more obvious. The military did not overtly intervene in civilian politics (though military agencies were used in covert ways by the ruling party) but was instead used to legitimise the regime. In particular, the OSRH's crucial role in the Homeland War directly linked it to the nationalist legitimacy of the Croatian state itself. By associating itself with the military and penetrating in the various ways described above, the Tuctman regime helped to bolster its own nationalist legitimacy in society.[55] The defence sector also continued to be staffed by HDZ loyalists and disagreements between military and civilian authorities were resolved by the dismissal of those who expressed opinions divergent with those of the regime. High profile examples of this strategy included General Petar Stipetić and Chief of the General Staff, General Anton Tus.[56]

The Tuctman regime also used the military to generate powerful lobby groups to support its policies. The most powerful of these groups was the Association of Veterans and Invalids of the Homeland War (HVIDRA). Throughout the late 1990s, HVIDRA consistently defended the regime whenever the international community criticised it. Its vocal opposition to the international communities' demands and strong nationalist rhetoric helped to legitimise the HDZ's refusal to cooperate fully with the ICTY. Partly through HVIDRA, the government persuaded a large portion of Croatian society to distrust the ICTY and to view the indictment of Croats on war crimes charges as illegitimate.[57] Until recently, fear of opposition from these groups put significant brake on the Račan government's cooperation with the ICTY and for a time appeared to threaten the whole process of Croatian integration into European institutions.[58] Since the HDZ regained power in 2003, groups like HVIDRA have become noticeably quieter. This is despite the fact that the Sanader government has pushed ahead with its predecessor's policies in a number of sensitive areas. Nonetheless, the veterans' associations continue to represent an important conservative element of the Croatian political discourse and are reflective of a wider societal ambiguity over some of the more negative implications of Croatia's war legacy.[59]

The incoming government in 2000 therefore faced a monumental task in reversing a decade of politicisation. The new defence minister, Jozo Radoš insisted that 'we intend to discontinue the practice of political activity within the Croatian Army... officers and civil servants within the Ministry of Defence will be allowed to belong to political parties but not to hold party functions. Our aim is to have experts in key positions in the Ministry of Defence'.[60] The minister then began a purge of political appointees from the ministry, though this proved to be much slower and more problematic than many observers had predicted. Similarly, the new President, Stipe Mesić, sacked twelve generals when they wrote an open letter to Croatian newspapers condemning the new government's policy of cooperation with the ICTY.[61] The government rescinded many of the President's powers and brought the defence sector under parliamentary control. However, as the previous section demonstrated, the divided nature of the Račan coalition led to a number of tensions between the government, presidency and parliament. The result, on the one hand, was institutional confusion. On the other, fractious coalition politics restricted the level of democratic scrutiny and on occasions even led to the replication of the Tuctman regime's politicisation strategy. For example, although Mesić won the presidency on a platform of removing the President's powers he has been very reluctant to rescind his authority over the armed forces. As Commander-in-Chief, he imposed his preferred choice as Chief of Staff, Petar Stipetić without consulting either the Defence Minister or Prime Minister. Some months later, officials in the Defence Ministry close to the minister tried to undermine the appointment by leaking stories that Stipetić had been implicated in war crimes by the ICTY.[62]

Building Professional Armed Forces

Because of the EU arms embargo, Croatian thinking on defence management was heavily informed by the United States. In 1994, confronted by the general embargo and acting on the advice of the US State Department, the Croatian government approached MPRI, a military consultancy firm run by a group of retired American Generals. MPRI was contracted to provide military training and education, which included a programme on the democratic control of armed forces and the subsequent Long Range Management Programme. Partly as a consequence of this influence, the Croatian defence establishment initially subscribed to a rather narrow understanding of military professionalism focused around all-volunteer forces. Moreover, they understood a 'good professional' to be one who volunteered for service and performed well in war at a tactical rather than somebody with the necessary expertise to fulfil the demands of a modern armed force.[63]

This perception is changing, though it remains predominant throughout much of the defence sector. An alternative view, subscribed to by the current Defence Ministry, is that professionalism and professionalisation are functions of education.[64] As a result, the current reform agenda in Croatia focuses on three key areas of professionalisation as a way of improving the capacity of the Croatian armed forces: first, the development of new roles to compliment tasks increasingly adopted by Western armed forces; second, the development of education and training programmes; and finally reforms to recruitment and promotion procedures. As with the reform of the control of the armed forces, each of these 'professionalisation' reforms has encountered significant problems produced by the environmental constraints described above.

New Roles

Because Croatia's armed forces were built in time of a war of national survival it is unsurprising that between 1991 and 2000 their role was seen in terms of territorial defence. However, two key factors have prompted civilian and military leaders to include provisions for new military roles within the reform process. First, political leaders have recognised the fact that military cooperation with other states and organisations can deliver significant political dividends and can help the government to achieve its primary foreign policy goals of NATO and EU membership. Second, the declining saliency of a direct military threat to national territory has forced defence planners in Croatia to ask serious questions about how best the OSRH can be reformed to address the country's new security environment. In doing so, they are in the process of reassessing the military's traditional defence of national territory role through a new (and at the time of writing, ongoing) Strategic Defence Review (SDR) process. This is developing new missions for the OSRH, including peacekeeping and humanitarian assistance, defence diplomacy, combating organised crime and providing domestic military assistance in times of crisis.

Peacekeeping in particular is viewed as an increasingly important new mission for the OSRH. Peacekeeping provides the most obvious form of international legitimation, conferring the legitimacy of the UN and other participating states. Additionally, it is widely thought that participation alongside NATO member states in peacekeeping operations will assist in moves towards interoperability and ultimately membership of the Alliance.[65] As part of this effort, the OSRH have contributed to a number of different peacekeeping operations. These include sending a military police platoon to NATO's ISAF mission in Afghanistan, and military observers to the UN missions in Sierra Leone, Ethiopia and Eritrea, Kashmir, Western Sahara, East Timor and Liberia.[66] Croatia has also signed a Stand-By Arrangement with the UN which commits it to have 40 military observers, 20 Staff Officers, 10 civilian policemen and seven military specialists available for deployment at any time.[67] Nevertheless, it is likely that it will be some time before Croatia is able to offer anything other than the smallest of contributions to multinational peacekeeping operations. This is largely due to the generally low standard of education within the military and financial limitations. Peacekeeping training is therefore conducted on a small scale with elite elements, contributing to 'patchy professionalisation'. Because of financial constraints, training is only offered to the small number of soldiers earmarked for peacekeeping duties.[68]

By contributing forces to multinational operations, Croatia both fulfils its obligations under its MAP, and illustrates that it is willing and able to contribute to the collective tasks and responsibilities of the Alliance more widely. Increasingly, the Ministry of Defence sees the role of the OSRH rooted in a European system of collective defence where direct military threats to national territory are unlikely but where sub-state and transnational security threats are significant. This is reflected in the current defence planning process which recognises that Croatia simply cannot afford to try and provide for full national defence capabilities on its own. Indeed, former Defence Minister Radoš went so far as to declare that 'in the case of an attack by a modern army, the Republic of Croatia will not be able to defend its sovereignty. We have neither the strength nor a modern army capable of doing that'.[69] As a consequence, capabilities are being restructured to recognise economic realities, with, for example, the Croatian air force's mission being limited in practice to patrolling the country's airspace in peacetime rather than aspiring to develop the capability to defend national territory in time of war.[70]

In moving away from the defence of national territory mission, defence planners in the Ministry of Defence are also working to tailor the OSRH towards the more immediate security challenges faced by Croatia today. These include, in particular, an increased role in providing assistance to domestic civilian institutions in their efforts to combat terrorism, trafficking and natural disasters.[71] At present, however, and beyond the downsizing process described below, it remains unclear the extent to which the structure, training and organisation of the OSRH has been changed to reflect these developments.

Education and Training

Croatian military education and training in the 1990s was influenced by two factors: the experience of war and the Tuctman regime's attitude towards the military. Because the HV was born in time of war, educational standards or military skill levels were not criteria that were taken into consideration when recruiting and promoting soldiers. Many people joined the armed forces without any formal civil education let alone military education and training. As with all armed forces during wartime, promotion was based upon tactical ability in the field rather than education, training, or other attributes.[72] Unlike the usual practice in other armed forces, however, these promotions were recognised after the war, creating an overly large officer cadre with a high proportion of officers having no military training and education or civilian education beyond the most basic level. Whilst those who were trained officers from the JNA did have such skills, their education was based upon the precepts of self-management socialism and the management techniques of the general peoples defence rather than the principles of war that ought to guide modern and professional armed forces.[73]

What military education there was provided after 1994 under the guidance of MPRI's democracy transition assistance programme. This involved an ad hoc collection of workshops, seminars, and courses that ranged in duration from a couple of hours to up to seven weeks.[74] This military training system did produce officers and soldiers capable of using basic equipment and solving tactical problems, as the OSRH's operational successes against the Croatian Serbs in 1995 testified. Officers did a particular course once they attained a particular rank in the field. However, these courses did not impart new skills or ideas nor did they have any impact on an officer's career.[75] One of the key problems confronting reformers in this area today is therefore a lack of indigenous civilian and military expertise.

To date, the reform of education and training for the OSRH has primarily been based on the utilisation of foreign education packages for Croatian officers. As well as the American International Military Education Training (IMET) programme for example, the US has organised numerous events, 'designed to present the US armed forces as a role model of a capable military under effective civilian control'.[76] More recently, several European states have also begun to offer assistance. The joint US/German centre at the Marshall Centre in Garmisch, Germany, has received over fifty students from Croatia on its courses on foreign and security policy management. Germany also accepts Croatian students at its military schools (for tactical level training) and its command and staff college (for higher level education). The UK has also been active in this regard, providing teacher training to three English language schools as well as several English-language courses in the UK that are vital if Croatia is to reach the required levels of interoperability for participation in PfP and peacekeeping.[77] Turkey, Italy, Hungary, Poland, France, Norway and Spain have also assisted in the provision of training and education either through direct activities or indirect support.

Access to foreign military education has undoubtedly provided the OSRH with a valuable external mechanism for improving the skill levels of its officers. However, a number of capacity and control problems remain that prevent this being exploited to its full potential. In particular, the lack of a developed human resource management system within the military or Ministry of Defence means that even when an officer has received military training abroad there is no systematic way of recognising these qualifications or linking them to job assignments or promotions. Perhaps more worryingly, some sources suggest that foreign educated students can be seen as a threat by conservative superiors and even risk being sidelined for promotion when they return to their units.[78]

Any long term solution to the OSRH's education and training needs must ultimately be provided indigenously. However, the key constraint on developing such as system remains funding. Whilst limited international funds for educational projects are available through programmes such as PHARE, it is unlikely that the finance needed to such wholesale reform could be forthcoming in the near future. In the meantime, interim measures using the facilities and expertise in the University of Zagreb alongside experts in the Ministries of Defence and Foreign Affairs have been adopted but tend to be rather ad hoc and are contributing to the process of 'patchy professionalisation' identified earlier.

Personnel

The issues of recruitment, retention, promotion, and the size of the armed forces are all interrelated. The Croatian armed forces have been, by general consensus, overstaffed. Indeed, in 2000, the total size of the armed forces was around 55,000 personnel, of which around 20–24,000 were conscripts.[79] The large size of the OSRH has created a number of problems. First, it has meant that a disproportionate amount of the defence budget has been allocated to personnel (see above, pp. 78–79, for details), leaving little money left over for investment in other areas. Second, it prompted the government to introduce a freeze on new recruitment into the military as a stop-gap measure while more comprehensive downsizing policies were developed and introduced. This in turn led to a situation where the average age of a soldier in the OSRH today is 36 years old.[80]

The government has tried to address these problems in two main ways. First, through a reduction of the length of national service, from ten to six months. This has allowed for four smaller intakes of draftees per year and reduced the number of conscripts in the armed forces at any one time significantly.[81] In addition, the Ministry of Defence is also looking at abolishing conscription altogether. The advantages of this is that there would be a quick personnel reduction in the OSRH together with an immediate (though modest) financial saving. More negatively, eliminating conscription would threaten many of the wider social and nation building roles that the OSRH currently play, as well as increasing the number of young unemployed in a very visible and way.[82] Conscription also remains a cheap way of providing military manpower and over the longer term developing and sustaining the alternative of a fully professional, all-volunteer force is likely to be an expensive option.

The second strategy for tackling the OSRH's large size and high personnel costs has been a more targeted programme of downsizing. Despite an almost universal acceptance within the Račan government that effective downsizing was crucial to the future development and professionalisation of the military, its implementation was initially slow. In the context of Croatia's high unemployment rate, many feared that reducing the size of the armed forces significantly would be tantamount to throwing thousands of 'war heros' on to the streets. The vocal opposition of groups like HVIDRA and the HDZ on the issue merely intensified the problem. Indeed, as late as the summer of 2002, there was no agreed strategy for how the downsizing programme would be implemented or how those selected to be 'downsized' would be chosen. This was despite the adoption of the document 'Decision on the Size, Composition and Mobilisational Deployment of the CAF' in May 2002.[83] This situation began to change later in the year when Radoš was replaced as Defence Minister by the more proactive and politically influential Zeljka Antunović.[84] Under Antunović, downsizing in the military was finally put into practice. This occurred the basis of voluntary redundancies, and in the context of a generous resettlement package for those who opted to leave.

On the surface, at least, Croatia's downsizing programme has been a great success, reducing the strength of the OSRH to 29,020 in January 2004. The new HDZ government has also continued with the same programme and aims to eventually bring the size of the military down to an ultimate figure of between 16–18,000 people.[85] Nevertheless, the programme has not been without its critics. Many argue for example that the cost of the programme has been too high and created even more of a bias towards personnel costs in the defence budget. Others suggest that the way the downsizing programme has been carried out—providing financial rewards for those that voluntarily choose to leave—encouraged only the brightest and best personnel to go, as these were the ones most constrained by the present system and for whom life outside the military would present the best opportunities.[86]

The need to reduce for downsizing has also been intimately linked to the dysfunctional structures of command created by the haphazard system of promotions put in place by the Tuctman regime. Under Tuctman, promotion within the military effectively came under the authority of either the President or Defence Minister Šušak. This was awarded either in return for political support or for tactical success in the field. Because of this idiosyncratic promotion system and the war, Croatia's armed forces became decidedly top-heavy. Since the Račan government came to power in 2000, responsibility for promotion within the military has also been the subject of struggles for influence between the government, parliament, presidency and general staff. The government is currently revising its policy on promotions though it has yet to find a consensus that can be implemented.[87]

Croatian armed forces have only recently begun their process of professionalisation, but some patterns are already emerging that reflect the broader impediments to reform described earlier. While there has been a consensus across the defence community about the need for transformation, economic, political, social and institutional constraints have hampered actual policy implementation. In particular, the Croatian defence sector has been faced with the tasks of responding to a massively changed geostrategic and domestic political environment that has placed new demands on the OSRH while simultaneously creating pressure for swinging cuts in the defence budget. Despite a number of fits and starts, both Račan and Sanader governments have attempted to square this circle by pursuing two main strategies. On the one hand they have pressed ahead with spending cuts and the downsizing programme with the aim of creating a more stable and sustainable basis for the long-term development of the OSRH. On the other, they have concentrated reform efforts in those areas which will deliver the greatest political returns, especially in relation to the key foreign policy goals of NATO and EU accession. This is most visible in the development of an active peacekeeping mission for the OSRH. Nevertheless, neither of these strategies is without risk. The former carries with it the danger of gutting the armed forces and loosing the highest calibre personnel in the process. The latter risks concentrating reform efforts on particular elite cadres and units within the military, encouraging 'patchy professionalistion' and stalling a more holistic approach to reform.

Conclusion

The Croatian case demonstrates both the potential for reform and the way that significant impediments can inhibit the reform process. It is clear that whilst the international community has played a significant role in helping to set the normative agenda for this process, there have been two main problems with the way it has attempted to assist the reform process. First, there is a strong correlation between elements of the defence sector that have been subjected to rapid and successful reform and elements that have been singled out for attention by foreign donors. More often than not such elements (such as peacekeeping) have not been singled out because of their importance to Croatia but rather because of their importance to foreign donors. As a result, areas of greatest need (such as provisions for conscripts, wholesale military and non-military education, equipment procurement) continue to be neglected. Second, the international community has shown little sensitivity to the structural impediments to reform described earlier. This has resulted in unrealistic expectations, and in the past, at least, the articulation of inappropriate and ineffective policies.

Croatia today finds itself at the cutting edge of a series of defence reform dilemmas. The current government broadly agrees with prevalent thinking in the international defence community and exhibits a strong desire to join Western European security institutions and to reform the defence sector in order to accomplish these wider policy goals. However, there are a series of competing demands that threaten this reform process. Croatia therefore confronts a series of paradoxes. First, the need to implement important defence reforms in the context of a rather unwieldy set of civil–military relationships, political and institutional rivalries, a lack of civil and military defence expertise and a continuing legacy of politicisation. Second, the need to cut defence spending as a proportion of the overall budget whilst taking on new military roles and improving the capability of the armed forces. Third, the need to balance the mission and interoperability requirements of the NATO accession process while implementing a balanced and fundamental reform of the armed forces as a whole. Finally, the need to implement root and branch personnel reforms and downsizing in the OSRH while simultaneously recruiting and retaining quality personnel and addressing the wider social issue of unemployment.

The success of Croatia's defence reforms will be heavily influenced by their relationship with wider political, social, and economic issues, and the role of external actors. If the international community is serious about assisting Croatia's reform process, and thereby enhancing regional peace and security and wider political transitions, it needs to adopt a more sensitive approach that recognises both the government's very real aspirations in this area, but also the significant impediments to reform that it confronts.

Notes

[1] Misha Glenny, *BBC Europe Online*, 1 August 2003.
[2] For further discussion of a typology of military roles see Timothy Edmunds, Anthony Forster & Andrew Cottey, 'Armed Forces and Society: A Framework for Analysis', in Anthony Forster, Timothy Edmunds & Andrew Cottey (eds), *Soldiers and Societies in Postcommunist Europe: Legitimacy and Change* (Houndmills: Palgrave-Macmillan, 2003). On the *nation builder* role: pp. 11–12.
[3] See Ozren Žunec, 'Democracy in the "Fog of War": Civil–military Relations in Croatia', in C. Danopoulos & D. Zirker (eds), *Civil–military Relations in the Soviet and Yugoslav Successor States* (Boulder: Westview, 1996, pp. 219–220); Alex J. Bellamy, 'A Crisis of Legitimacy: the Military and Society in Croatia', in Forster *et al. Soldiers and Societies*, pp. 186–188.
[4] Many of these had received rudimentary military training as members of Croatia's network of Territorial Defence (TO) units or as conscripts in the JNA. Zvonimir Mahečić, 'Capability-Building and Good Governance in Security Sector Reform', in Jan A. Trapans & Philipp H. Fluri (eds), *Defence and Security Sector Governance and Reform in South Eastern Europe: Insights and Perspectives* (Geneva: Geneva Centre for the Democratic Control of Armed Forces, 2003), pp. 404.
[5] Biljana Vankovska, 'Privatisation of Security and Security Sector Reform in Croatia', in Damian Lilly & Michael von Tangen Page, *Security Sector Reform: The Challenges and Opportunities of the Privatisation of Security* (London: International Alert, 2002), pp. 56–57.
[6] For a more detailed discussion of the government's role in propagating a particular form of national identity see Alex J. Bellamy, *The Formation of Croatian National Identity: A Centuries-old Dream?* (Manchester: Manchester University Press, 2003).
[7] Žunec, 'Democracy in the "Fog of War"', p. 225. One analyst went so far as to call the OSRH under Tuctman 'one of the most politicised militaries in the postcommunist world'. Biljana Vankovska, 'Military and Society in War-torn Balkan Countries: Lessons for the Security Sector Reform', in Alan Bryden & Philipp Fluri (eds), *Security Sector Reform: Institutions, Society and Good Governance* (Baden-Baden: Nomos Verlagsgesellschaft, 2003), p. 231.
[8] Alex J. Bellamy, 'Like Drunken Geese in the Fog: Developing Democratic Control of Armed Forces in Croatia', in Andrew Cottey, Timothy Edmunds & Anthony Forster (eds), *Democratic Control of the Military in Postcommunist Europe: Guarding the Guards* (Houndmills: Palgrave, 2002), 178–180.
[9] Many suspect that the arms embargo was not wholly adhered to by all European countries. Germany in particular has been accused of turning a blind eye to illegal exports of East German weaponry to Croatia as well as assisting in the establishment of the Croatian intelligence services. Tim Ripley, 'Croatia's Strategic Situation', *Jane's Intelligence Review* (1 January 1995); Marko Milivojević, 'Croatia's Intelligence Services', *Jane's Intelligence Review* (1 September 1994).
[10] For more on MPRI and its relationship with the US government see Vankovska, 'Privatisation', p. 65, 69–71.
[11] These are based on defence spending figures provided by HIDRA, the Croatian Information and Documentation Agency. We are grateful to Renata Pekorari for providing this information.

[12] Lidija Cehuić, 'Development of Civil–military Relations in Croatia', *International Issues*, 10: 1 (2001), p. 116.
[13] This legacy was still apparent as late as May 2002 when civilian employees in the Ministry of Defence and General Staff numbered 9,183 out of a total peacetime force of 42923. *Defence Policy 2004/05: Into the Alliance* (Zagreb: Ministry of Defence of the Republic of Croatia, January 2004), p. 8.
[14] The role of the military in the 2000 election is outlined in more depth in see Alex J. Bellamy, 'Croatia after Tuctman: The 2000 Parliamentary and Presidential Elections', *Problems of Post-Communism*, 48/5 (2001), pp. 18–31.
[15] In contrast to the nationalist and authoritarian Tuctman-era HDZ party, the HDZ under Sanader has made a considerable shift towards the political mainstream. While many suspect that the party retains many of its more unpalatable nationalist tendencies at grassroots level, the HDZ in government can best be characterised as a centre-right party with strongly pro-EU and pro-NATO leanings. See Edin Forto, 'Croatia', in Adrian Karatnycky, Alexander Motyl & Amanda Schnetzer (eds), *Nations in Transit 2004* (New York and Washington: Freedom House, 2004), p. 5.
[16] Croatia joined PfP in May 2000.
[17] Timothy Edmunds, *Defence Reform in Croatia and Serbia–Montenegro*, Adelphi Paper 360 (Oxford: Oxford University Press, 2003). pp. 16–17; Jelena Grčić-Polić, 'Security and Defence Reforms: A Croatian Armed Forces Case', *Croatian International Relations Review*, 9: 30/31 (2003), p. 18.
[18] See for example, Zvonimir Mahečić, 'Civilians and the Military in Security Sector Reform', in Trapans & Fluri, *Defence and Security Sector Governance*, p. 376; Richard B. Liebl, Marin Braovać & Adrijana Jelić, 'Security Assistance Programmes: The Catalyst for Transition in the Croatian Military', *The DISAM Journal* (Spring 2002), pp. 7–8.
[19] Robin Luckham, Democratic Strategies for Security in Transition and Conflict', in Gavin Cawthra & Robin Luckham (eds), *Governing Insecurity: Democratic Control of Military and Security Establishments in Transitional Democracies* (London: Zed Books, 2003), pp. 13–18.
[20] Jane Chanaa, 'Security Sector Reform: Issues, Challenges and Prospects', *Adelphi Paper No. 344* (Oxford: Oxford University Press, 2002), p. 28.
[21] Zvonimir Mahečić, 'Security Sector Reform in Croatia', in Timothy Edmunds (ed.), *Security Sector Reform in Croatia and Yugoslavia*, DCAF/IISS workshop proceedings, 28 October 2002, p. 15.
[22] The Prime Minister, Ivica Račan was a reformist communist/socialist and the president, Stipe Mesić is a moderate nationalist. See Forto, 'Croatia', pp. 4–8.
[23] Edmunds, *Defence Reform*, pp. 16–17
[24] Edmunds, *Defence Reform*, pp. 18–20.
[25] Edmunds' interviews, Zagreb July 2004. This is an accusation the government itself denies, pointing in particular to its plans to introduce a fairer and more transparent human resource management system for the Ministry of Defence.
[26] Parliament's role in this area is also hampered by secrecy laws which restrict parliamentarian's access to classified documents. Vlatko Cvrtila, 'Parliament and the Security Sector', in Trapans & Fluri, *Defence and Security Sector Governance*, pp. 369–370.
[27] Two examples include the Institute for International Relations (IMO) at http://www.imo.hr and the Centre for Defendological Research (CDR) at http://www.defimi.hr
[28] See Anthony Forster, Timothy Edmunds & Andrew Cottey, 'Introduction: The Professionalisation of Armed Forces in Postcommunist Europe', in Anthony Forster, Timothy Edmunds & Andrew Cottey (eds), *The Challenge of Military Reform in Postcommunist Europe: Building Professional Armed Forces* (Houndmills: Palgrave Macmillan, 2002), p. 6.
[29] See for example, Andrew Cottey, Timothy Edmunds & Anthony Forster, 'The Second Generation Problematic: Rethinking Democracy and Civil–military Relations', *Armed Forces and Society*, 29: 1 (2002), pp. 41–44.

30 Bellamy, 'Like Drunken Geese', pp. 178–180. Dimitrios Koukourdinos, 'Constitutional Law and the External Limits of the Legal Framing of DCAF: The Case of Croatia and the FR of Yugoslavia', in Biljana Vankovska, *Legal Framing of the Democratic Control of Armed Forces and the Security Sector: Norms and Reality/ies* (Geneva: Geneva Centre for the Democratic Control of Armed Forces, 2001), pp. 157–158.

31 Marina Ottaway & Gideon Maltz, 'Croatia's Second Transition and the International Community', *Current History,* November (2001), pp. 379–380.

32 Mahečić, 'Civilians', p. 375.

33 Damir Grubiša, 'Democratic Control of Armed Forces' and Cvrtila, 'Parliament' in Trapans & Fluri, *Defence and Security Sector Governance*, pp. 351, 356, 357, 365.

34 Edmunds' interview with Zvonimir Mahečić, Assistant Head, Military Cabinet of the President of the Republic of Croatia, 11 June 2002.

35 Edmunds *et al.*, 'Armed Forces and Society', pp. 9–14.

36 Cehulić, 'Development of Civil–military Relations', p. 121.

37 One observer noted wryly in 2004 that while the Račan government were democrats that had to prove their nationalist credentials, the HDZ are nationalists who have to prove their democratic credentials. Edmunds' interview, Zagreb, July 2004.

38 Bellamy's interviews with Zdenko Franić, Social Democratic Party management committee member, 12 May 1998 and Božo Kovačević, Liberal Party politician and former government minister, 14 May 1998. See also Eric Kopač, 'Economic Constraints of Defence Reform in SEE', *Croatian International Relations Review*, 8: 26/27 (2002), p. 34.

39 See for example, Zoran Kusovac, 'HDZ Tests Croat Coallition', *Jane's Intelligence Review*, 13: 4 (April 2001).

40 For example, the latest national security documentation stresses repeatedly that Croatia faces no direct military threat from any direction. *Defence Policy 2004/05*, p. 11.

41 In his election campaign, Nikolić made specific reference to his dream of a 'Greater Serbia' whose expanded borders would include significant parts of Croatian territory. See 'Nikolic Still Dreaming of a Greater Serbia', *B92 News Archive*, 23 June 2004. Available at http://www.b92.

42 Chanaa, 'Security Sector Reform', p. 30.

43 Discussed further in Alex J. Bellamy, 'Like Drunken Geese'.

44 The Croatian defence budget fell from 6.084 bn Kuna or 4.4% of GDP in 1998 to *Defence Policy 2004/05*, p. 19; Kopač, p. 33.

45 Jozo Radoš, 'Statement to the IMO–NATO Conference on Regional Stability and Cooperation: NATO, Croatia and South-Eastern Europe, Zagreb, 24–25 June 2002', *Croatian International Relations Review: Dossier*, 8: 26–27 (2002), pp. 14.

46 See for example the policy priorities identified in *Defence Policy 2004/05*, pp. 17–18.

47 Vankovska, 'Privatisation', p. 69–71.

48 Edmunds' interview with Polić, 7 July 2003. See also Amadeo Watkins, *PfP Integration: Croatia and Serbia-Montenegro*, CSRC Balkans Series Report 04/5 (Camberley: CSRC, April 2004), p. 8.

49 Ozren Žunec, 'Democracy in the "Fog of War"', p. 216.

50 Žunec, 'Democracy in the "Fog of War"', p. 220; Milivojević, 'Croatia's Intelligence Services'.

51 Žunec, 'Democracy in the "Fog of War"', pp. 224–225.

52 Edmunds, *Defence Reform*, p. 14.

53 Vankovska-Cvetkovska, 'Between the Past and the Future', p. 40.

54 Ozren Žunec, 'Democratic Oversight and Control Over Intelligence and Security Agencies', in Trapans & Fluri, *Defence and Security Sector Governance*, p. 384; John Hatzadony, *The Croatian Intelligence Community*, Federation of American Scientists, Intelligence Resource Programme Report (1999). Available at http://www.fas.org/irp/world/croatia/hatzadony.html (accessed 15 March 2005).

55 Bellamy, 'A Crisis of Legitimacy', pp. 193–194.

56 Žunec, 'Democracy in the "Fog of War"', p. 224.

[57] For further discussion of the role of HVIDRA see, Kusovac, 'HDZ Tests Croat Coalition', p. 21; Bellamy, 'Like Drunken Geese', p. 181; Edmunds, *Defence Reform*, pp. 18–19.

[58] See in particular events surrounding the attempted extradition of General Janko Bobetko to the Hague. Edmunds, *Defence Reform*, pp. 62–63.

[59] Vankovska, 'Military and Society', pp. 233–235.

[60] Kusovac, 'Interview with Jozo Radoš'.

[61] Kusovac, 'HDZ Tests Croat Coalition', p. 21.

[62] Bellamy, interview with senior official from the Croatian Ministry of Defence, 14 October 2001.

[63] For a discussion of military professionalisation see: Forster *et al.*, 'Introduction'.

[64] Edmunds' interview with Dr Jelena Grčić-Polić, Assistant Minister for Defence Policy, Republic of Croatia, 07 July 2004; Edmunds interview with Brigadier Željko Cepanec, Head, Defence Policy and Planning Department, Ministry of Defence of the Republic of Croatia, 7 July 2004.

[65] Edmunds' interview with Zoran Milanović, Head of the Department for International Security, Ministry of Foreign Affairs of the Republic of Croatia, 19 July 2002.

[66] *Defence Policy 2004/05*, p. 9.

[67] Zvonimir Mahečić, 'Peacekeeping and Regional Security', in Trapans & Fluri, *Defence and Security Sector Governance*, p. 460.

[68] Mahečić, 'Peacekeeping and Regional Security', pp. 459–460.

[69] *Slobodna Dalmacija*, 22 March 2002. Cited in Ozren Žunec, 'Croatia's Decision to Abandon the upgrade of Mig-21 Aircraft with the Israeli Company', unpublished Draft Paper, 2004, p. 25.

[70] Žunec, 'Croatia's Decision', pp. 25–26.

[71] *SDR: CAF Missions and Tasks (Draft)*, information provided to Edmunds by the Ministry of Defence of the Republic of Croatia, 7 July 2004.

[72] Wartime Defence Minister Šušak, for example, observed that 'war experience is much more important than some diplomas'. Quoted in Cehulić, 'Civil–military Relations', p. 127. See also, Zvonimir Mahečić, 'Capacity-Building and Good Governance in Security and Defence Reform' in Trapans & Fluri. *Defence and Security Sector Governance*, pp. 404–406.

[73] On the basic concepts that underpinned the JNA approach to war fighting, such as the general peoples defence, see Adam Roberts, *Nations in Arms: The Theory and Practice of Territorial Defence* (London: Chatto & Windus, 1976), pp. 124–217 and the opening chapters of James Gow, *Legitimacy and the Military: The Yugoslav Crisis* (London: Macmillan, 1992).

[74] Alex J. Bellamy, 'A Revolution in Civil–military Affairs: The Professionalisation of Croatia's Armed Forces', in Forster *et al.* *The Challenge of Military Reform*, p. 173.

[75] Mahečić, 'Capacity-Building', p. 405.

[76] Wheaton, 'Cultivating Croatia's Military', p. 11; See also, Liebl *et al.* 'Security Assistance Programs'.

[77] Edmunds' interview with Lt.Col. Richard Thornely, UK Defence Attaché, Zagreb, 08 July 2002.

[78] Edmunds' interview, Zagreb, 6 July 2004.

[79] Bellamy, 'A Revolution', p. 175.

[80] *Defence Policy 2004/05*, p. 18.

[81] From 12,041 in May 2002 to 6,885 in January 2004. *Defence Policy 2004/05*, p. 8.

[82] Croatia's unemployment rate is very high, currently standing at around 19% of the population in 2003. 'Croatia', *CIA World Factbook 2004*, Available at http://www.cia.gov/cia/publications/factbook/ (accessed July 2005).

[83] Edmunds, *Defence Reform*, pp. 40–43. Watkins, *PfP Integration*, p. 8.

[84] In contast to Radoš, Antunović was also a Deputy Prime Minister and had enough political clout to force the policy through.

[85] *Defence Policy 2004/05*, p. 8. Edmunds' interview with Polić, 7 July 2004.

[86] Edmunds interviews, Zagreb, July 2004.

[87] For a more detailed discussion of some of the problems and complexities of the promotion system in the OSRH see Grubiša, 'Democratic Control' and Mahečić, 'Capacity-Building' in Trapans & Fluri, *Defence and Security Sector Governance*.

The Transformation of Romanian Civil–military Relations: Enabling Force Projection

LARRY L. WATTS
Assistant to the President of the Romanian Senate, Oversight Committee for Defence, National Security and Public Order, Bucharest, Romania

The Romanian civil–military relationship remained on a remarkably even keel as it weathered the implementation of democratic civilian control and the transformation of the military from a mass-conscript territorial army to a more professional NATO-interoperable force capable of power projection. Throughout these changes, public support for the army and its new roles consistently remained the highest in the post-communist space. The relatively smoother path of civil–military relations during the transformation owed much to the independence of the Romanian Armed Forces (RAF) from Soviet military control before 1989.[1]

Beginning in early 1960s, Romanian officers were no longer trained in the USSR, the RAF stopped participating in Warsaw Pact troop exercises, and intelligence cooperation with the KGB and GRU was terminated.[2] As the only Warsaw Pact member to refuse participation in and publicly condemn the Soviet-led invasion of Czechoslovakia in 1968, Romania set itself against the pact and its international goals. By the late 1970s the country and its army were universally regarded as probable adversaries of the USSR in any East–West conflict.[3]

The popularity gained by the RAF as a result of this confrontation, along with the regime's dependence on the army for the protection of its other autonomous policies, was resented by communist dictator Nicolae Ceausescu. He treated the RAF with increasing suspicion and hostility, even while he exploited it as a source of cheap labour, further strengthening its identification with an increasingly repressed population.[4] The RAF's decision to side with the demonstrators in the December 1989 revolution, whereupon it suffered over 20% of the revolution's casualties, cemented this linkage and created levels of civilian respect and admiration for the military unmatched in Central and Eastern Europe.

The sudden and violent nature of the revolution overthrew all other state and government institutions and rid the RAF of communist party structures and personnel literally overnight.[5] In contrast, the negotiated transfers of power common among other former members of the pact resulted in compromises that preserved communist influence and continued to hamper military reform efforts as, for example, the separation of the General Staffs and army command from defence ministries did for over a decade in both Hungary and Poland.[6] Consequently, neither the population nor civilian leaders (whether government or opposition) viewed the military as a threat to the consolidation of democracy, and Romania's transition was unhindered by the pre-existing civilian mistrust of the military (and visa versa) common elsewhere in the postcommunist space.[7]

Having already performed the basic institution-building tasks facing the other transition armies, and extremely popular, the RAF was able to undertake the 'first generation' reform of basic structures and legal frameworks immediately after the revolution. It also had a firm basis to address 'second generation' reforms associated with the exercise of parliamentary and budgetary control, personnel advancement and reconversion, professionalization, and defence diplomacy and defence planning for new missions.[8]

While Romania avoided the common error of prematurely appointing inexpert civilians to leadership posts within the Ministry of Defence (MoD), the civil–military relationship did suffer crises generated by misdirected initiatives of civilian political authorities who inadvertently politicized the military during the first months of transition.[9] When the new political leadership reactivated an anti-Ceausescu general and appointed him defence minister in 1989, whereupon he reactivated 30 other superannuated officers in the highest posts, the army almost disintegrated.[10] The disaster narrowly averted, civilian and military authorities recognized that 'shared responsibility' was required to prevent similar crises, and they jointly developed a strategy to create a community of civilian defence experts and advance reform while preserving military effectiveness and the stability of civil–military relations.[11]

Two important early innovations that remain critical to the civil–military relationship were the Supreme Defence Council (*Consiliul Suprem Aparare de Tara*: CSAT), established in 1990, and the National Defence College, founded in 1991. The CSAT is the most important locus of security reform and defence planning, bringing together all executive authority in Romania's semi-presidential system where the presidency is primarily responsible for national security and foreign policy and the prime minister for domestic administration. Its members include the president (as chair), the prime minister, the presidential counsellor for national security, the directors of foreign and domestic intelligence, the chief of the General Staff, and the ministers for defence, foreign affairs, internal affairs, and industry. Decisions are made by consensus and are binding on all state and government institutions.

The CSAT created the National Defence College (NDC) to train senior politicians and opinion leaders alongside officers in order to foster a non-partisan, shared understanding of defence issues and security problems. This enabled civilian politicians from government and opposition, military leaders, and independent civilian analysts to 'read off the same page' much earlier than their regional counterparts.[12] All of these factors established an environment of confidence and dialogue permitting an arrangement whereby the military carried the water for civilian-designated reforms. As a result, most reform disputes remained intra-military and did not provoke civil–military tensions during the administrations of Ion Iliescu from 1989 until the end of 1996. Although civil–military relations and reform temporarily suffered during the 1997–2000 administration of Emil Constantinescu, their underlying stability was underscored by the quick repair of both after the change of administration in the elections of December 2000.[13]

Democratic Control of Armed Forces

Democratic control of the military was stipulated in decree laws before the first election in May 1990, and reaffirmed in a series of laws issued before and after the adoption of the constitution in 1991.[14] Romania has the full array of legislation necessary for control over the defence establishment in war and peace, with the exception of a law on states of siege and emergency, drafts of which have been hung up in parliament since the mid-1990s. Since January 2001, British, American and German military and civil–military advisers seconded to the MoD and General Staff have assisted and monitored the transformation first hand, attending, for example, the minister's regular planning meeting with the MoD leadership, the Chief of the General Staff (CGS) and the service chiefs.[15]

Parliament fulfils its oversight function through its Committees for Defence, National Security and Internal Order in both houses founded in 1990, by the president as commander-in-chief, and by the government through its defence minister. Although the Constitutional Court, Ombudsman, State Audit Court, and Ministry of Public Finance also have formal military oversight competencies, only the latter two are occupied with day-to-day oversight functions.

The president is constitutionally responsible for foreign policy, the defence of the country, and public order. His powers to declare partial and general mobilization, and states of siege and emergency, are tempered by the dual requirement of parliamentary and CSAT approval. According to the constitution, the presidency develops the fundamental documents on defence and national security policy, the most important of which is the *National Security Strategy*. This has been the practice since 1994 when the first strategy (then termed the *Integrated Concept of National Security*) was created. New strategies were introduced in 1999 and 2002. The strategy is prepared by the President's Department of National Security in close cooperation with the CSAT and the MoD, after which comments from the General Staff are invited.

Beginning in 1999 the government began issuing the *White Paper on National Security and Defence*, establishing the objectives, actions to be taken, and resources that the various ministries and agencies must provide in meeting the provisions of the *National Security Strategy*. The most recent *White Paper* was approved by the Romanian Parliament in May 2004 and takes into account the new status of Romania as a NATO member, providing for new missions and tasks to be fulfilled by the Romanian Armed Forces including: (1) contributing to the peacetime security of Romania; (2) defending Romania and its Allies; (3) promoting regional and global stability, including through the use of defence diplomacy; and (4) providing appropriate support to state and local authorities in case of civil emergencies.

Based on RAF missions and tasks, the *Military Strategy* establishes force structure and assigns resources to achieve the objectives set down in the *National Security Strategy* and the *White Paper*. Prior to 2001 the *Military Strategy* was formulated by the General Staff's Directorate for Strategic Planning to the exclusion of the civilian MoD. As of April 2001 the strategy is fully controlled by the MoD and the minister's Defence Planning Council.

Parliamentary Oversight and Transparency

Genuine parliamentary control requires civilian expertise that cannot be developed overnight. In Romania, this need was partly addressed by the attendance of committee members in the NDC and similar foreign programs since 1992. As of 2004 nine members of the parliamentary committees for defence oversight completed the NDC defence management program. Several others have attended programs at the George C. Marshall Centre and the Monterey Institute. The problem is also partly addressed by the established traffic between defence oversight committees and the ministries preoccupied with national security. For example, former defence ministers have served on the lower house oversight committee since 1996 and, beginning in 2005, the committee includes two former defence ministers, a former state secretary for defence planning, a former chief of the general staff, and former presidential counsellor for national security among its members.

The committees have broad powers for approving the defence budget and organizational changes, hearing civilian officials and officers, investigating problems and inspecting facilities. However, during the 1997–2000 Constantinescu administration the government often resorted to emergency ordinances that avoided parliament altogether.[16] Thus, a 'core goal' of the defence minister in 2000–2004 was 'the re-empowerment of the parliamentary authority over the military and defence policy'.[17] A deputy defence minister (state secretary) post for parliamentary relations was created, a civilian appointed, and adequate funds allocated. The new Department for Parliamentary Relations, Legislative Harmonization and Public Relations had two offices for relations with the senate and the chamber of deputies responsible for constantly informing them of military developments. It also had two offices for harmonizing legislation with NATO and the EU, an office for relations with non-governmental organizations, and an office for relations with the public and mass media.

In 2001–2004 the defence committees met at least once and often 3 or 4 times a week during parliamentary session, usually with the MoD state secretary for parliamentary relations in attendance. Their members served on no other committees.[18] When military reorganization measures were at issue, the minister or a state secretary was obliged to participate. Although Romania's committees have been among the most active in their oversight functions for some time, until 2001 they were deficient in the single most important area of democratic control: oversight of the defence budget and expenditures.

Defence Planning and Budgetary Control

During 1990–3, MoD underwent several radical reorganizations such that regular planning processes were disrupted and carried on by more informal groupings of civilian and military leaders. Defence planning was produced largely by the General Staff and then discussed in the CSAT. More consistent civilian control was made possible with the creation of a Department for Defence Policy and International Relations and the appointment of a civilian state secretary to head it in March 1993. Defence policy during 1993–6 was the product of negotiated consensus undertaken within a civil–military framework where civilian leadership on the 'big' issues was uncontested. But it did not reflect consolidated civilian control over the building blocks of which defence policy was constructed.[19] A new national defence planning system was legislated in 2000 but could not be implemented until 2001 because of fundamental problems with the state budgeting process and tensions created by dysfunctions in the ruling coalition.

The heart of the problem was that defence planning and budgeting are inextricably linked. From 1989 through 2000, continued uncertainty at the level of state budgeting made medium and long-term budgeting in the military unachievable. Typically, the state budget was only passed during the fourth–sixth month of the year of its application and was means-based rather than program driven. This made real planning impossible and discretionary spending was the norm, leading to severe problems in procurement, interoperability, waste and corruption. In 1993 the Chief of the General Staff requested specialists from the Institute for Defence Analysis (IDA) to help the RAF introduce the Planning, Programming, Budgeting and Evaluation System (PPBES) used by the US Armed Forces. PPBES delegates responsibility for specific programs according to well-defined requirements, timescales and costs, thus enabling forward planning and greater accountability. Although the implementation of PPBES was thenceforth included as a reform goal, persistent uncertainty in the state budget and the lack of required expertise in program-driven budgeting blocked the comprehensive introduction of PPBES before 2001–2.

An agreement concluded in 1998 with the US Naval Undergraduate School for Defence Resources Management in Monterrey, California to establish a Romanian institution that would train civilian defence and military personnel to manage the PPBES was a major step in filling this gap, but the Democratic Convention government balked at funding the project. The Regional Centre for Defence Resource Management began training shortly after the Social Democrat (PSD) government allocated these funds in 2001.

Although the 2001 state budget was not adopted until April, the 2001 defence budget was the first to be program-driven, thus permitting more forward planning and greater accountability. Concurrently, the PPBES was implemented throughout the MOD, and extended throughout the military in 2002. The PPBES was run by the civilian chief of the Department for Defence Policy and Euro-Atlantic Integration, State Secretary George Maior, through the civilian-led Defence Integrated Planning Directorate (DIPD) in the MoD. Complementary planning and budget structures also function within each service.

Program-driven defence budgets for 2002, 2003 and 2004 were approved by the government and parliament in the autumn of the year prior to their application. For the first time since 1989 a complete inventory of military equipment was carried out during June–August 2001 and the first inventory of all real assets was completed in 2002. Since 2001 both the RAF and the parliamentary oversight committees have had the ability through PPBES to trace allocations to specific military units, their breakdown, and how they are actually expended. These advances created the resource transparency necessary for effective democratic control.

Aside from parliamentary scrutiny during the budgetary approval process, the military is subjected to audits by the State Audit Court, which reports directly to parliament and has a comptroller at the MoD full time. The Ministry of Public Finance also has two comptrollers in the MoD, and both they and the Audit Court comptroller must countersign all documents issued by the MoD's financial directorate.

The government formally established the Audit Directorate within the MoD in 1999, and the structures necessary to carry out its functions were created and made operational in 2001. This directorate monitors the legality of specific allocations and expenditures—a task also performed by the comptrollers and by the Inspectorate General—as well as the efficiency of those allocations and expenditures. It is also responsible for ensuring that MoD finances are handled in accord with EU standards.

Under the direction of the minister's Defence Planning Council (DPC), the state secretary and the DIPD are also responsible for the generation of the single joint defence plan and single joint defence budget. The DIPD was set up during the Constantinescu administration but entrenched General Staff autonomy over defence planning and budgeting (the initial director and staff of the DIPD were military officers) and the lack of governmental support for changing the status quo rendered it ineffective prior to 2001 when the current changes were initiated. These changes have since been enumerated in defence ministry ordinances and written into law by parliament.

The single defence plan is drawn up in the DIPD based on information and figures from the MoD staff and General Staff. It is then discussed and approved first in the DPC and finally by parliament. The DIPD works with the program managers in preparing the plan while the CGS usually endorses the programs developed by the single services, by the Logistics Command, and by the Signal Command. In the shared opinion of the NATO state advisers who sit in on the minister's weekly planning meetings, the Romanian MOD now has a joint planning system well in hand and under democratic civilian control, enabling them to forward plan five years and beyond.[20]

Media

Post-communist civilian and military elites were quick to grasp the fact that the public image of the armed forces is a crucial component of its combat capability. Any decrease in public support could diminish the military's ability to fulfil its basic missions. In order to build and maintain public support for reform at home and operations abroad, the MoD instituted a public outreach program under its Directorate of Public Relations (DPR) in 1991 that has made the military one of the most transparent institutions in Romania.

The DPR regularly informs the public on how public funds are spent, on the progress of reform efforts in downsizing and restructuring, and on military training, combat readiness, and military living standards. It also reports on the status of NATO integration efforts and on participation in UN, Partnership for Peace and NATO-led peace support and counter-terrorism operations. During 2001, for example, 79 civilian media organizations and 102 journalists were accredited to the MOD and the MOD issued 409 press communiqués, over four times more than any other ministry. Journalists are regularly included on working visits at home and abroad. The DPR monitors media coverage of the military by nine TV stations, three wire services, and 32 publications, responding to criticism and correcting inaccuracies.

It also produces a weekly newspaper, a weekly radio broadcast, and a weekly TV program. As part of its mandate the DPR organizes international conferences and public debates on topics such as 'The Armed Forces and Society', 'Military Careers and Youth in the Armed Forces', 'Transparency of Security—Security of Transparency', and 'Defence of Transparency—Transparency of Defence'. Since 2001, press briefing transcripts and recordings, backgrounders, and the Romanian Military Newsletter have also been made available to the media and the public via the Ministry's internet website (http://www.mapn.ro), more than doubling the public information activities of the RAF. The MOD worked with civilian universities to introduce a syllabus on NATO studies in 2001. It has also assisted in the formation of the civilian led NATO Studies Centre in Bucharest which is sponsored by NATO, the Romanian MOD and a university in Bucharest.

Civil Society

As an additional means of democratic control, the 'strategic community' within Romanian civil society is beginning to exert real influence. The process has been rather slow, partly due to the former practice of generating these institutions within the RAF and granting their analysts military rank prior to 1989. The most influential defence and strategic institutes so far have been those embedded in international associations, such as the Mannfred Woerner Euro-Atlantic Association and the George C. Marshall Association. These institutes also have the greatest degree of civilian–military cross-fertilization. In 2001 they joined forces with several other associations to form the NATO House (*Casa* NATO). Other institutions specializing in this domain, such as EURISC for example, have also begun to establish their own influence.

Aside from the common problems in developing communities of expertise, partial explanation for the delay is that prior to 1996 most Romanian NGO's representing civic society were 'captured' by opposition parties and became extremely partisan. For example, the overt partisanship of the Civic Alliance culminated with its entrance into the Democratic Convention government in 1997, while the Group for Social Dialogue campaigned openly for Constantinescu's re-election. Although they are still hobbled by overt partisanship, there has been a general maturing of defence and security NGOs over the last seven years, assisted by the large number of analysts that have passed through the training program for defence managers at Romania's National Defence College thus laying the ground work for common approaches to these issues.

Professionalization and Military Reform

Since 1989 the RAF has downsized its military and civilian manpower from 320,000 to 120,000 and its active service peacetime army from almost 290,000

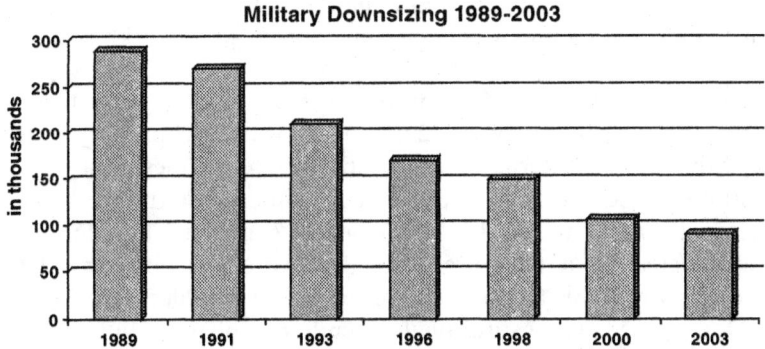

Figure 1. Military downsizing 1989–2003. *Source*: J-5, General Staff, Romanian Ministry of National Defence.

to 90.000 (see Figure 1). The peacetime force target of 75,000 to be reached in 2007 is on schedule. Since the military itself 'carried the water' for this restructuring, it did not create severe tension in military–society relations. Serious downsizing of the officer corps was undertaken beginning in 2001, when over 4,000 senior officers, including almost 1,000 colonels and over 400 generals, were made redundant. The process has continued with more than 2,000 senior officers selected to leave the military in 2003. Potential social issues arising from the need to reabsorb these personnel into civilian life were addressed through a comprehensive reconversion process undertaken with the assistance of the World Bank. The program provides for job placement, small business start-up counselling, and professional cross-over training. The Higher Military Academy and Staff College—now the National Defence University— also provided new career training for officers within 3 years of retirement and the MOD has post-retirement employment agreements with several government ministries.

The designers of Romania's defence and security reforms rapidly understood the implications of the post-Cold War security environment for military missions and structures, although it has taken additional time for some senior officers to reach a similar level of comprehension. Despite Romania's membership in the Warsaw Pact, its strategy prior to 1990 was classic territorial defence where a mass-conscript, heavily armed force deployed on the home territory to combat invaders alone. As new non-state threats replaced traditional military threats, emphasis shifted from organizing one country against a security threat to organizing for security cooperation within the NATO alliance.

The new conditions required the army to play a more active constabulary and diplomatic role entailing 'new requirements for force-building, flexibility, sustainability, and strategic deployment capabilities.'[21] All of this implied the demise of the territorial army and an end to conscription. Without a military threat territorial armies were little more than unskilled manpower, a budgetary drain, and a breeding ground for corruption, patronage networks, and declining morale.

Conscription was cut back periodically over the last ten years to less than 14,000 annually in a population of 22 million, of which the primary beneficiary was the Ministry of Internal Affairs and its gendarmes, not the MoD. Conscientious objectors have been able to choose alternative community service since 1997. Although the MOD openly supported the elimination of conscription, creating a 68% professional all-volunteer force by 2003, the interior ministry was able to maintain conscription until 2005.[22] In March 2002 the MOD proposed a constitutional amendment phasing out conscription which parliament approved in October 2003. The 2003 force-planning concept originally envisioned complete professionalization by 2007 but was accelerated by one year in 2005.

Personnel Policy

Along with radical downsizing, ensuring the political neutrality of the military was an immediate priority of the post-communist leadership. Membership in or activity by military personnel on behalf of political parties was outlawed in a December 1989 decree and this prohibition was restated in the 1990 MoD law, in the 1991 Constitution, in the 1994 National Defence Law on the Romanian Armed Forces, and in the 1995 Law on the Status of Military Personnel. True protection against politicization, however, requires an objective, fair and transparent human resource policy. Otherwise subjectivity and expediency can easily lead to preferential promotion policies based on political criteria, deprofessionalizing the military.

In Central and East Europe politicization morphed with the collapse of communism from the use of military force by the regime for internal defence against the population to the enlistment of the military by ruling governments in the domestic partisan political competition. The most common avenue for this sort of politicization is through the political manipulation of officer career patterns, creating a corps of officers who owe posts and promotions that they would not have achieved either at all or so quickly through their own merit and effort to a particular government.

Before 1989, personnel management was the task of the Higher Political Directorate (HPD) and already highly subjective because Ceausescu's hostility provoked a defensive mechanism whereby officers' evaluations were routinely rated 'very good', rendering the system of limited use for judging actual merit and performance. When the HPD was dissolved in the revolution the responsibility for promotions and redundancy decisions fell to the CGS, relying on the advice of his subordinates. This increased the subjectivity of the process since officers tended to favour personal loyalty in the absence of objective criteria. The problem was exacerbated in that a host of officer promotions were granted immediately after the revolution to those deemed to have suffered politically-motivated career limitations under communism.[23] In many cases these promotions were also based on haphazard and extremely subjective criteria. In all cases they were made without regard to concrete restructuring goals.

The 1995 Law on the Status of Military Personnel establishing regulations for military promotion marked the first real step in personnel reform. However, the selection process remained the responsibility of a small group of officers whose procedures and criteria were non-transparent and therefore suspect. In 1997 a Human Resources Department was established to advise the CGS but selection continued to be highly subjective and the department exercised overly broad discretion in interpreting promotion guidelines. Moreover, a loophole allowing 'promotion by exception' existed in the military personnel law through which one could drive a truck.[24] For example, CGS Constantin Degeratu, appointed by President Constantinescu in 1997, was eligible for the position only because of a series of exceptional promotions that brought him from lieutenant colonel in 1995 to a four-star general on his retirement at the beginning of 2001, while his successor, Mircea Chelaru, rose from major in 1993 to 3-star general by 2000 on the basis of the same exceptional process. The damage could only be repaired by creating and implementing a credible, transparent, and fair system of personnel selection and retention. There was common agreement between the presidency and the new defence minister and his staff that real civilian control over the military would not be possible until this problem was brought under control. As Minister Pascu noted on his appointment, the new system had to:

> Eradicate present politicising practices; establish strong mechanisms for preventing any future re-politicisation, and reinstate professionalism as the main criterion for career advancement. [It must also] reconfigure the command corps in line with the projected reduction of forces, in accord with the current plans of downsizing and modernisation. Specifically, the Western system will be adopted in which, even if one has the time-in-rank, experience, and appropriate exam scores for promotion, he or she will be compelled to retire from active service if there are no slots available.[25]

The implementation of the new personnel management system benefited significantly from British assistance and the determination of Pascu's team, led by state secretary Maior, and was accomplished in several phases. First, the law regarding the status of military personnel was modified to meet the personnel needs of the military and to protect the rights of officers and enlisted personnel in accordance with Western standards. Second, a *Military Career Guide* describing the new system and laying out the guidelines for a promotion process based on position requirements and individual qualifications was approved and published.[26] The *Guide* stipulated that all promotions to and within the officer corps were to be done by designated military selection boards, setting forth their organization and scope of activity.[27] The composition and competence of the selection boards were further detailed in a ministerial ordinance.

The first promotion board met in July 2001. Outside reviews conducted in the summer of 2002 and 2003 continued to improve the process, bringing the RAF's personnel policies closer to those of NATO member states.[28] Beginning in July 2002, British, American and German advisers periodically observe the board proceedings during their deliberation. Military personnel are now ranked and promoted on the basis of proficiency and potential, and positions are correlated to rank. There is a mandatory retirement age of 55. The transparency of the RAF's personnel management system managed to dampen much of the political backlash associated with officer downsizing.

Under the 1990–6 Iliescu administrations the MoD employed senior and mid-level civilians from both government and opposition parties. During the Constantinescu administration in 1997–2000 all civilian opposition party members were dismissed from the MoD, undermining what had been an effective working relationship between civilian defence managers and military officers. In order to repair the damage and create a stable pool of civilian expertise within the MoD, the new administration restricted civilian political appointments to the level of minister and state secretary, enforced the civil service law, and initiated the development of a career guide for civilian MoD employees, thus addressing a problem that plagued most of the new NATO members.[29]

The Military and Society

The RAF is popularly associated with every significant advance in state-building and national consolidation since the formation of the Romanian people. It began the post-communist transition with enviably high levels of public trust—upwards of 90%—and has consistently received the trust of over three-quarters of the population ever since.[30] The shift in the basis for the army's legitimacy has not lessened public support as regional instabilities in Romania's corner of Europe and threats that arose after the Cold War have invested both traditional and new normative roles with greater currency. The former Soviet military threat was replaced by the threat of 'grey zone' insecurity should Romania remain outside the NATO alliance.

Defence Diplomacy

The RAF thus aggressively engaged in defence diplomacy and peace-keeping operations because of necessity and its capacity to do so. The success of these efforts, for example, in the 1990 Open Skies agreement with Hungary at a time when the two governments were barely on speaking terms, bolstered the RAF's reputation as a pragmatic institution advancing general national rather than partisan or particular interests.

Romania's slogan of being a 'regional security producer rather than consumer', coined in 1993, reprised a role it played in the inter-war period when it co-founded the Little Entente. To this end, Romania has initiated, co-founded, or participates in the Romanian–Hungarian Joint Peacekeeping Battalion; in the *Tisza* Engineer Battalion; in the UN's Stand-by High Readiness Brigade (SHIRBRIG); in the Black Sea Naval Task Group (BLACKSEAFOR); in the Central European Nations' Cooperation Initiative (CENCOOP); and has been chair of the South Eastern Europe Security Cooperation Group (SEEGROUP) which ensures connectivity between NATO's South Eastern Europe Initiative (SEEI) and regional cooperative initiatives.

Romania chairs the Political and Military Steering Committee for the Multinational Peacekeeping Force South Eastern Europe (MPFSEE) with the USA, Italy, Greece, Turkey, Bulgaria, Slovenia and Albania. It also contributes an infantry battalion, reconnaissance platoon, transport platoon, engineer company, and staff personnel to the South East European Brigade (SEEBRIG), based in the Romanian port of Constanta. State secretary Maior, especially, was active in developing a Black Sea strategy to cope with the troublesome issues of the extended Black Sea region before President Traian Basescu, elected in 2004, made it a national priority.

Force Projection

Since 1989 Romania has deployed more than 14,000 military personnel—more than 14% of its current active service peacetime army—for various humanitarian, peace support and combat operations in the Middle East, Europe, Africa, and Central Asia. Its decision to act as a de facto ally of NATO in September 2001 was unanimously approved by parliament and wide-ranging, offering troops for the global war against terrorism wherever they might be needed. It has stood by this commitment. The RAF doubled its military presence in the Balkans to ease redeployments to Afghanistan and US KFOR forces rotated out of Kosovo through Romanian port and air facilities on the Black Sea.

In January 2002 Romania was the first NATO candidate to commit and then deploy forces (military police, air transport and staff officer contingents) in Afghanistan under the British-led International Security Assistance Force. In July, the RAF deployed an infantry battalion for combat missions as part of the US-led *Operation Enduring Freedom* and has maintained a consistent force level by rotating additional infantry battalions on that operation. In March 2005 it committed an additional 400 troops to OEF.

In April 2003, the RAF deployed units with US forces as part of the coalition effort to oust Saddam Hussein. Several months later it deployed an infantry battalion with the Italian Brigade in the British-led Multinational Division and

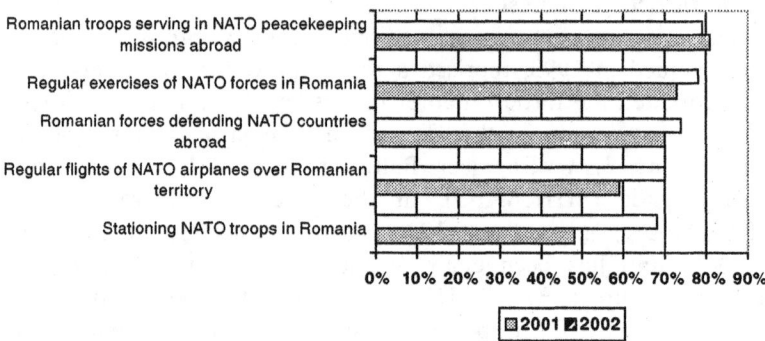

Figure 2. Public support for military roles in NATO.

military police, intelligence units and an engineering detachment with Polish-led forces in Iraq. An additional 100 troops were designated for Iraq in March 2005. Romanian forces in Afghanistan and Iraq were transported by air and sea largely with national means—a feat many long-standing NATO members could not achieve. In 2003 and 2004 the RAF was deploying and logistically supporting more than 1,700 personnel in peace support and combat operations in Afghanistan, Bosnia, Iraq and Kosovo.

According to polls in 2001 and 2002 (see Figure 2), popular support for the military's new normative roles, particularly under NATO auspices, is robust for both peacekeeping and war-fighting.[31] Support has increased despite the eleven casualties the RAF incurred in Afghanistan, Angola, Bosnia and Kosovo. General regional instability undoubtedly plays a role in generating this support. Missions that bolster regional security directly serve Romania's national security. There is, of course, broad acceptance of the norms and values that underpin these new roles. Contribution to NATO missions also strengthens the alliance Romanians depend upon for their own security. The country's larger role in international peacekeeping and enforcement further grants it a respectable 'seat at the table' in discussions regarding the future security of the region and of Europe, while bolstering its credibility as a legitimate partner.

From the military reform perspective, participation in these roles provides an efficient method of improving interoperability and compatibility with NATO faster, as well as providing the experience necessary for constituting a fully professional force, two goals which enjoy strong public support. Additional incentives include the increased opportunities for advancement for personnel with peacekeeping, crisis management and combat experience and pay rates many times greater than at home. The fact that the troops have consistently performed well in NATO, EU, UN and OSCE missions also constitutes a point of pride for all Romanians.

Influences on Changing Civil–military Relations

NATO has easily been the most influential factor in the transformation of civil–military relations. Once Romania decided in 1993 that becoming a member of the alliance was a national security priority of the first order, NATO pressure and assistance complemented and reinforced domestic pressures for internal reform and integration. In part, this reflects the popular desire to reintegrate into the West after a half-century of isolation. It is also an artefact of the security and prosperity that the organization, along with the EU, represents for the general population.

In 1990, reform was broadly conceived as seeking to achieve modern 'Western-like' military structures to enable possible cooperation with the other side of the former Iron Curtain. By early 1991, the goal was refined to creating a military that could cooperate with NATO forces. After 1993, when NATO announced its intention to open the alliance to new members, the primary goal of reform was to create a NATO military that would be welcomed into the Alliance.

At this point, military reform and NATO integration became a unified process. Once NATO's various Partnership for Peace (PfP) and Membership Action Plan (MAP) initiatives were centralized in 2001, real assessments of where the reform process stood and how to move forward were generated.[32] This translated directly into more open communication and greater transparency such that both Romanian officials and NATO gained detailed knowledge of what obstacles were still to be overcome and NATO state advisers were able to tailor their recommendations and advice to address specific problems with much greater effect. This in turn created a 'sea change' in Romanian military reform and integration in terms of pace and effectiveness. The MAP process ensured that promised reform was monitored and that much lauded plans were actually implemented. In 2001 Romania focused on 13 key objectives and in 2002 was able to demonstrate that actual progress had been made in all of them. In real terms it produced the evidence of progress requested by NATO rather than the endless recital of plans not implemented which characterised the 1996–2000 period.

All military reforms within the RAF are considered on the basis of their compatibility with NATO standards and practices and overseen by a special Directorate for Military Reform and NATO Integration. The NATO MAP process, in particular, proved enormously beneficial for setting priorities, monitoring fulfilment, and ensuring consistent practice.[33] Although not directly concerned with the military, the EU integration process, and particularly the accession advisers placed within various ministries, also proved a boon to military reform by establishing a framework of reforming institutions and a template of European legislation and practice.

Romanian support for NATO integration has consistently been the highest in central and eastern Europe at around 85 to 87%. During the Kosovo campaign support for NATO dropped to an all-time low of 56%, partly due to Romanian scepticism regarding the utility of the operation for resolving ethnic conflict, and partly to the proximity of the bombing to the Romanian border.[34] However, even this diminished support was on a par with that of new NATO member Poland (at 60%), equal to that of Germany and France, and exceeded support levels in Hungary, the Czech Republic and Greece.[35]

Social Change

Its unusual status as a fully field-capable military at the start of the democratic transition and the lack of inherited civil–military antagonisms enabled the RAF to undertake a number of measures in restructuring and addressing social issues that led developments within society at large. For example, the recruitment of women, frozen after the fall of communism, was reintroduced in 2001, and they now make up 3% of the officer and NCO corps.[36] Selection criteria are similar to that for male personnel and regulations provide equal educational, training and career opportunities. In 2004, there were over 1300 female officers and NCOs, with some 200 women in the Higher Military College, almost 100 in NCO schools, and more than 50 in the various service academies.

The ethnic Hungarian party in parliament occasionally calls for proportional representation in the officer corps but the military career has not proven attractive to young ethnic Hungarians despite recruitment drives by the MoD, probably due to the still relatively low living standard of military personnel.[37] This may change since expectations of improvement in the military standard of living experienced an upswing beginning in 2001.[38]

The RAF introduced tolerance training (largely through the British–Romanian Regional Training Centre for Staff Officers) several years ago and is currently developing a tolerance/diversity training program with US and British assistance at the basic training and NCO levels. Although the Roma/Gypsy population is the main target group of such programs, they also address the problem of minority recruitment and retention generally. A Roma commission has been set up and is operating within the RAF to work on recruitment issues and to address specific problems of military personnel from this community. In March 2002, the National Defence College introduced a Holocaust teaching unit in its program in cooperation with the US Holocaust Museum.[39]

Economic Constraints

Romanians generally are very supportive of the downsizing and professionalization of the RAF.[40] In spite of the economic exigencies, and contrary to what might be surmised by popular support for a smaller professional military, the Romanian armed forces are in the enviable situation of also enjoying broad public support for larger defence budgets. According to a 2001 poll, a significant majority of the population (63%) favoured increased defence allocations even at the expense of other public sectors, indicating that Romanians neither equated 'smaller' with 'cheaper', nor (currently) baulked at further economic trade-offs in order to ensure their security in a relatively unstable region.[41] With the single exception of 1999, defence has received 2% or more of the state budget, with significant annual increases since 2000.[42] Given the steady improvement of the economy since 2000 and the prospects over the next five years, overall economic pressure on the RAF is more likely to remain constant than suffer any significant increase that would adversely influence current policies and the civil–military relationship.

However, making the hard choices on where to focus scarce resources is a permanent issue during modernization and transition. Once it chose to create a smaller, mobile, flexible and interoperable force capable of power projection it actively engaged in PSO and other missions as an instrument for modernizing and professionalizing the armed forces, while at the same time employing a 'training the trainers' strategy—using the experience gained to modify its reforms and improve its operational performance. For example, the head of J7 Training and Doctrine at the general staff was the commander of the brigade whose troops were deployed in Afghanistan and Iraq. The central choice was not one of 'reform or perform' but rather what reforms best enhance mission performance and how best to use performance to refine reform goals and accelerate the modernization process. This has been possible largely because of the dialogue between the MoD and General Staff on the one hand, and between the MoD and the CSAT on the other.

Conclusion

The RAF is a professional military under democratic control in both the broad and narrow sense—firmly subordinated to civilian authority, capable of power projection missions, and more than two-thirds all-volunteer. According to the US Ambassador to NATO, Romania was already playing a 'leading role in NATO' even prior to its formal accession to the Alliance in April 2004.[43] The lack of prior civil–military antagonisms and a common civil–military understanding of the necessity for radical restructuring in the post-Cold War environment is one element that made this possible.

Another important element in the relationship is the continuing contribution to military legitimacy made by the traditional and new normative roles, by on-going regional instability, by new threats of terrorism and organized crime, by the regularity of seasonal environmental disasters requiring military assistance, and by the RAF's status as the leading institutional reformer. The army's impressive performance in the NATO integration process and on missions abroad constitute the third leg of Romania's robust civil–military relationship.

Paradoxically, rifts within Europe between NATO members over the appropriate role of European states in NATO and *vis-à-vis* the United States have the greatest potential for disturbing this relationship, particularly when they create pressure to 'choose between' NATO and close relations with the US on the one side and the EU on the other. Romanians, most of all those who have served in the military, see NATO and the EU as two sides of the same coin of Western re-integration—NATO for security and defence and the EU for social policy and economic prosperity. They view both as unitary in terms of a common set of democratic norms and values. Consequently they have shown obstinate dedication to both while rejecting pressures for partisan short-sightedness that, among other things, has the potential for complicating the newly-transformed civil–military relationship.

Notes

[1] In contrast: 'At no point in the communist period was the Polish General Staff in a position to plan for an all-out defence of the country's territory', and the Polish Defence Ministry 'had no control over the Polish operational army'. Andrew A. Michta, *The Soldier–Citizen: The Politics of the Polish Army after Communism* (New York: St. Martin's Press, 1998), p. 84.

[2] Czech and Slovak officers, for example, continued to attend Soviet military academies and institutes until the demise of the USSR (66 attending in the summer of 1991, with 8 more slated for training in 1992). Thomas S. Szayna and James B. Steinberg, *Civil–military Relations and National Security Thinking in Czechoslovakia* (Santa Monica: RAND, 1992), p. 18. See also, Alex Alexiev, *Romania and the Warsaw Pact: The Defence Policy of a Reluctant Ally* P6270 (Santa Monica: RAND Corporation, 1979), p. 17.

[3] See e.g., A. Ross Johnson, *The Warsaw Pact: Soviet Military Policy in Eastern Europe* P 6583 (Santa Monica: RAND Corporation, 1981), 4n, 19, pp. 30–4; Ryszard J. Kuklinski, 'The War Against the Nation as Seen from Inside', *Kultura* (Paris), 4: 475 (Spring 1987), pp. 3–57; and Victor Suvorov, *Inside the Soviet Army* (New York: Berkeley Books, 1982), pp. 3–11.

[4] Walter M. Bacon, Jr., 'Romanian Military Policy in the 1980s', in Daniel N. Nelson (ed.), *Romania in the 1980s* (Boulder: Westview, 1981), pp. 209–10.

[5] All state and government bodies, with the exception of the army, were dissolved by decree of the provisional leadership on 25 December 1989.

[6] For example, Decree no. 8 on Political Parties from 31 December 1989 forbade any political activity within the military and any military personnel from membership in a party.

[7] Christopher Donnelly, 'Developing a National Strategy for the Transformation of the Defence Establishment in Post-Communist States', *European Security*, 5:1 (Spring 1996), p. 5.

[8] According to British analysts, 'Romania's unique military position within the Warsaw Pact ensured that the military were able to think and take decisions for themselves'. *Review of Parliamentary Oversight of the Romanian Ministry of National Defence and the Democratic Control of its Armed Forces*, DMCS Study No. 43/96 (London: Directorate of Management and Consultancy Services, 1997), p. 30. In contrast, US analysts concluded that 'high-ranking Polish officers never made major decisions'. Michta, *The Soldier-Citizen*, p. 101.

[9] Douglas L. Bland, 'Protecting the Military from Civilian Control: A Neglected Aspect of Civil–military Relations', in *Democratic and Civil Control Over the Armed Forces: Case Studies and Perspectives* (Rome: NATO Defence College International Symposium, 1994), pp. 107–127. Larry L. Watts, 'Reforming Civil–military Relations in Post-Communist States: Civil Control vs. Democratic Control', *Journal of Political and Military Sociology*, 30:1 (Summer 2002), pp. 51–70.

[10] Larry L. Watts, 'The Romanian Army in the December Revolution and Beyond', in Daniel Nelson (ed.), *Romania After Tyranny* (Boulder, Westview, 1992), pp. 95–126.

¹¹ See Douglas L. Bland, 'A Unified Theory of Civil–military Relations', *Armed Forces and Society*, 26:1 (fall 1999), pp. 7–26.
¹² The success of the NDC prompted the CSAT to create the National Security College in 2001 in an effort to create an equally non-partisan understanding of security and intelligence issues among civilian politicians and journalists.
¹³ Larry L. Watts, 'The Crisis in Romanian Civil–military Relations', *Problems of Post-Communism* 27:4 (July/August 2001), pp. 14–26.
¹⁴ Law no. 39/1990 on the Establishment, Organization and Functioning of the Supreme Council of National Defence; Law no. 41/1990 on the Organization and Functioning of the Ministry of National Defence; Law no. 45/1994 on the National Defence of Romania; Law no. 73/1995 on the Preparation of the National Economy and Territory for Defence; Law no. 80/1995 on the Status of Military Personnel; Law no. 46/1996 on the Preparation of the Population for Defence; and Law no. 106/1996 on Civil Protection.
¹⁵ Coupled with the progress reports and feedback of NATO's Partnership for Peace and Membership Action Plan, these advisers made it possible to ascertain the real state of affairs instead of relying on Romania's highly partisan and notoriously unreliable press. See e.g., International Federation of Journalists, *Money, Power and Standards: Regulation and Self-Regulation in South-East European Journalism—Practices and Procedures in Albania, Bulgaria, Croatia and Romania* (Brussels: Royaumont Process Project, November 1999), pp. 10–12, 26–32.
¹⁶ Almost 1,000 ordinances failed to reach parliament before the administration was voted out of office in 1996.
¹⁷ Ion Mircea Pascu, 'Immediate Defence Priorities', Bucharest, December 2000, p. 3.
¹⁸ Foreign intelligence committee members are drawn from the defence committees but meetings of both are scheduled at the same time.
¹⁹ Larry L. Watts, 'Reform and Crisis in Romanian Civil–military Relations: 1989–1999', *Armed Forces and Society*, 27: 2 (Summer 2001), pp. 597–622.
²⁰ Author's conversations with General Richard Smith, Ms. Jacqueline Callcut, General (ret.) Ed Buckley and Colonel Rudiger Volk during 2001–3.
²¹ George Cristian Maior and Mihaela Matei, 'Bridging the Gap in Civil–military Relations in Southeastern Europe: Romania's Defence-Planning Case', *Mediterranean Quarterly* 14: 2 (Spring 2003), p. 66.
²² *Comunicat M.Ap.N. nr.55/8 feb—Precizari SMG privind executarea serviciului militar*, available at http://www.mapn.ro. Alin Bogdan, 'Desi Constitutia desinfinteaza, Ministrul Rus vrea sa pastreze stagiul militar obligatoriu la pompieri si jandarmi', *Adevarul*, 3 September 2003.
²³ George Cristian Maior, 'Personnel Management and Reconversion' in Larry L. Watts (ed.), *Romanian Military Reform and NATO Integration* (Iasi–Oxford: Center for Romanian Studies, 2002), pp. 61–2.
²⁴ The various means of 'promotion by exception' are noted in articles 53, 63 and 64 of Law no. 80/1995 on the Status of Military Personnel.
²⁵ Pascu, 'Immediate Defence Priorities', p. 4.
²⁶ Human Resources Management Directorate, *Military Career Guide* (Bucharest: Romanian Ministry of National Defence, 2001).
²⁷ *Military Career Guide*, Chapter V. Articles 51–53.
²⁸ In contrast, the Polish president appoints all flag officers whereas the Romanian president confirms the appointments decided by the military selection boards and approved by the parliamentary defence committees. See e.g., Elizabeth Coughlan, 'Democratisation and the Military in Poland: Establishing Democratic Civilian Control', in Constantine P. Danopoulos & Daniel Zirker (eds), *The Army and Society in the Former East Bloc* (Boulder: Westview, 1999), p. 125.
²⁹ Tomas Zipfel, 'The Politics and Finance of Civil–military Reform in the Czech Republic', in David Betz and John Lowenhardt (eds), *Army and State in Postcommunist Europe* (London: Frank Kass, 2001), p. 106.
³⁰ See, e.g., www.state.gov/www/background_notes/Romania/0700_bgn.html.

[31] Metro Media Transylvania, *Population's Attitudes Concerning the Military*, February 2001 and March 2002.
[32] Jeffrey Simon, 'Roadmap to NATO Accession: Preparing for Membership', *Institute of National Strategic Studies (INSS) Special Report*, National Defence University, October 2001, pp. 1–8.
[33] See, e.g., Larry L. Watts, 'Democratic Civil Control of the Military in Romania: An Assessment as of October 2001', in Graeme P. Herd (ed.), *Civil–military Relations in Post Cold War Europe* (Sandhurst: Conflict Studies Research Centre, December 2001), pp. 14–42.
[34] Metro Media Transylvania, May 1999.
[35] Public opinion polls for NATO performed by Metro Media Transylvania for March 1997, May 1999, May 2000, February 2001. The December 1998 poll performed by the Centre for Urban and Rural Sociology (CURS).
[36] Major General Constantin Gheorghe, 'Force Restructuring', in Watts, *Romanian Military Reform*, p. 129.
[37] See, e.g., *Adevarul* and *Curierul National*, 4 March 2002.
[38] Psycho-Social Research Section, General Staff, *Values, Norms and Mentalities in the Military Environment* (Bucharest: Romanian Ministry of National Defence, 2001).
[39] *Radio Free Europe/RL NEWSLINE* 6:52, Part II, 19 March 2002.
[40] Metro Media Transylvania, *Population's Attitudes*, February 2001.
[41] Metro Media Transylvania, February 2001.
[42] In 1998 defence expenditures dropped to 1.77%. Allocations in 2002 and 2003 amounted to 2.38% of the GDP.
[43] Victor Roncea, 'Romania este deja un lider in NATO', *Ziua*, 31 August 2003.

Civil–military Relations in Serbia–Montenegro: An Army in Search of a State

TIMOTHY EDMUNDS
Department of Politics, University of Bristol, Bristol, UK

The electoral defeat of Slobodan Milošević and his Serbian Socialist Party in October 2000 marked a turning point for the Federal Republic of Yugoslavia (FRJ)[2]. The new government—formed by the Democratic Opposition of Serbia (DOS) coalition—appeared to offer the country a pathway out of the isolation and ethnic nationalism of the 1990s. It also offered the prospect of a transformation in the country's troubled civil–military relations. Under Milošević, these had been strained and characterised by mutual suspicion and military politicisation. Despite these tensions, the Yugoslav Army (VJ)[3] had ultimately been co-opted by the regime and during the October 2000 system change many feared that it would intervene in its defence. Organisationally, the VJ was shaped by the experience of conflict and international isolation. It was oversized and much of its equipment and infrastructure was either obsolete, warn out or war damaged. Moreover its reputation and professionalism had been tarnished by its role in the dissolution of Yugoslavia and its association with successive military defeats and war crimes. In 2000 therefore, the FRJ faced a threefold challenge in relation to reforming its civil–military relations: first, establishing the principle of civilian control over the armed forces and democratising the nature of this control; second implementing a comprehensive programme of military modernisation and organisation reform; finally, managing an evolving and sometimes troubled relationship with the West—particularly over the issue of war crimes responsibility and cooperation with the International Criminal Tribunal for the Former Yugoslavia (ICTY).[4]

Since the fall of Milošević, FRJ/Serbia–Montenegro (SCG) has taken a number of major steps forward in these areas: the principle of civilian control over the military has been established; there has been a clear public recognition from across the defence sector—including both military and civil elements—that civil–military reform and modernisation are necessary and urgent; a number of important personnel changes have taken place; the Army and the Ministry of Defence have both undergone limited reorganisation; the government has also entered into a dialogue with NATO over potential membership of the Partnership for Peace (PfP) and initiated some limited cooperation with the ICTY. These reforms have taken place in the context of an extremely negative set of legacies from the 1990s. These include the experience of war, authoritarianism and nationalism, as well as international isolation and ultimately NATO bombing in 1999. As such, their significance should not be underestimated or downplayed.

Nevertheless, civil–military reform in what is now SCG has not lived up to the more optimistic expectations of October 2000. In practice it remains closely linked to—and ultimately dependent on—wider political and societal processes of post-conflict and post-authoritarian transformation. These have been dogged by two interlinked challenges. First was the need to fashion a strong and stable governmental programme for reform. This has been made more difficult by the deeply fractured nature of the domestic political scene, the underlying conservatism of the military itself, the persistence of strong nationalist sentiment in Serbian society as a whole, and continuing western pressure and conditionality on the ICTY issue. Second was the need to conduct reform in the context of a continuing crisis of legitimacy and authority in the State Union itself and open questions over the future independent status of Montenegro. The impact of these challenges on civil–military reform in SCG has largely been negative. At best they have led to stagnation and delay in the reform process. At worst they have threatened the repoliticisation of the armed forces and placed the future of military reform in the country at the mercy of partisan political struggles in the civilian sector.

Civil–military Relations under Milošević

Civil–military relations in SCG today remain deeply influenced by the legacy of the Milošević period. This has imparted a legacy of civilian control over the armed forces, but also of persistent civil–military distrust and politicisation.

At the beginning of the 1990s, the then Yugoslav People's Army (JNA) was not a natural ally of Milošević. Although its officer corps had always been dominated by ethnic Serbs, the JNA had fulfilled an explicitly *pan-Yugoslav* rather than pro-Serbian role in the old Socialist Federative Republic of Yugoslavia (SFRJ).[5] The JNA also had a strong tradition of professionalism and institutional autonomy which made it resistant to manipulation by the Milošević regime.[6] Despite these legacies, as the Yugoslav crisis of the late 1980s and early 1990s progressed the JNA found itself increasingly drawn into the Serbian nationalist project, and by Spring 1991 was actively fighting on its behalf. Nevertheless, Milošević never wholly trusted the VJ and throughout the 1990s attempted utilised a number of strategies to subjugate it to his own control.

The first of these was the establishment of new chains of command that bypassed or manipulated the FRJ's existing constitutional provisions for civil–military relations. Under the SFRJ's 1974 constitution, command responsibility over the Army lay in the hands of the federal presidency. Given that for most of the 1990s Milošević was Serbian rather than federal president, this arrangement would have given him no formal authority over the military. This changed in 1992, however, with the introduction of a new a new constitution and a new chain of command for the VJ. This established a Supreme Defence Council (VSO) comprising the federal, Serbian and Montenegrin presidents, which would exercise ultimate command over the military.[7] In practice, this gave Milošević, as both the Serbian and subsequently federal president, a new and constitutionally sanctioned mechanism for exercising direct influence over the VJ.

Second, Milošević worked to change the political and ethnic character of the military itself through personnel changes and purges in the officer corps. This process began in the early 1990s with the resignation or early retirement of many non-Serbs from the JNA.[8] Throughout 1992–3, many of the remaining JNA officers in the VJ were either dismissed or forced into early retirement in a series of purges which left only nine former JNA generals in the whole institution, and eliminated any vestiges of old-style 'Yugoslavism' in the new VJ.[9] Even then, the Army continued to exhibit what for Milošević was a worrying degree of autonomy and it was only in 1998, when the firm Milošević loyalist General Dragoljub Ojdanić was appointed Chief of the General Staff, that he fully consolidated his personal control over the VJ's leadership.[10]

Finally, Milošević undermined the influence of the VJ through the cultivation of direct institutional competitors from other parts of the Yugoslav security sector. This strategy included the militarisation of the Serbian police forces and the development of a number of semi-autonomous special forces and paramilitary organisations outside the Army's normal chain of command. These alternative military formations forced the VJ to compete for favour and funding from the regime in Belgrade, and functioned as an alternative power-base for Milošević should the Army prove unreliable.[11]

Milošević's policies towards the Army have had a lasting impact and continue to problematise the SCG civil–military relationship across both civil and military sectors. In particular, they have left a legacy of legislative ambiguity and a tradition of security sector politicisation that have hampered the consolidation of democratic control over the armed forces since 2000.

Democratic Control of Armed Forces

Establishing Civilian Control

The first civil–military challenge faced by the new DOS government was a stark one: how to prevent the VJ (and indeed other elements of the FRJ security sector such as police special forces) from intervening to prevent the system change—and then subsequently how to keep it out of domestic politics and under firm civilian control. This was far from an abstract concern. The VJ leadership was packed with Milošević cronies whose own positions and interests appeared endangered by any political change and who had publicly implied that they would defend the regime by force.[12] Moreover, any process of democratisation or rapprochement with the West would most likely raise uncomfortable issues of war crimes responsibility and root and branch organisational reform. This threatened to expose the Army to greater external scrutiny and challenge its still privileged and relatively autonomous position within the FRJ.

DOS's response was to negotiate with the leadership of both the military and the police in the hope of persuading them not to intervene in Milošević's defence. This strategy was significant for two reasons. First, it suggested the emergence of a new pattern of security sector politicisation in FRJ that mirrored the deep personal and political divisions within the DOS. These focused around the two largest political parties and personalities in the coalition: Vojislav Koštunica and his conservative nationalist Democratic Party of Serbia (DSS) and Zoran Đinđić and his more reform minded Democratic Party (DS).[13] Koštunica therefore used the October negotiations to form the basis of a close relationship with the VJ and its controversial Chief of Staff General Nebojša Pavković. Đinđić did likewise with the Serbian police.[14] In one sense this division of power and influence was uncontroversial: Koštunica as federal president was de facto Commander in Chief of the federal Army, while Đinđić had command of the republican police forces. However, given the rivalries within the DOS—and indeed between the VJ and the Serbian police themselves—this arrangement also provided both Koštunica and Đinđić with their own potentially partisan supporters and power bases in the FRJ security sector. Second, some interpretations of the events of this period suggest that the police and Army's non-intervention in the October 2000 system change was bought with the promise of a 'soft' approach to military and police reform (and particularly personnel reform) by the new government.[15] Certainly, at least one key DOS politician from the time has admitted that the new government was nervous of pushing too hard on security sector reform too early for fear of sparking a praetorian backlash.[16] This in turn suggested that the transition to democracy in FRJ was one that would have to be 'negotiated' with the country's security sector—at least in certain sensitive areas. It also suggested that the extent to which the military was actually under firm civilian control in practice remained open to question.

In the event, the VJ—in common with the police and FRJ's other paramilitary and security services—did not intervene to prevent Milošević's fall. Nevertheless the civil–military legacy of October 2000 was a largely negative one that left the extent to which the Army was actually under civil control—rather than simply tolerating it temporarily—open to question. Certainly, for much of 2000–2 Koštunica appeared reluctant to push through reforms in the Army, particularly in relation to the dismissal of Milošević appointees in its leadership. Of these, General Pavković was perhaps the most significant. Pavković had been a staunch Milošević loyalist and as commander of the VJ's Third Army in Kosovo was implicated in ethnic cleansing and war crimes.[17] Throughout 2000–2 he regularly commented on sensitive issues of domestic politics in FRJ—such as cooperation with the ICTY—in an open an partisan manner.[18] Pavković also engaged in a bitter and public dispute with former Chief of General Staff and Serbian Deputy Prime Minister Momčilo Perišić, which culminated in the latter's arrest by the military Counter Intelligence Service (KOS) on a charge of spying in March 2002.[19] Koštunica's reluctance to move on Pavković appears to have been informed by his rivalry with Đinđić and was indicative of the continued politicisation of FRJ's civil–military relations. In return for a tentative approach to military reform and protection from the attentions of the ICTY, Pavković provided Koštunica with a useful—if unpredictable—ally at a time when Đinđić's power in the DOS was in the ascendant.[20]

By June 2002, domestic and international pressure on Koštunica to dismiss Pavković had reached its peak. Further civil–military reform appeared unlikely whilst he remained in position. His public comments undermined the government's efforts to place the Army under civilian control, while NATO had countries made it clear that they would not do business with the VJ while he was still in place. This stance had direct ramifications for FRJ's evolving relationship with the Alliance and particularly its potential membership of PfP.[21] Koštunica eventually moved to dismiss his controversial Chief of Staff in March 2002, soon after the Perišić affair. He did not manage to do so successfully until late June, after a convoluted and divisive series of events that exposed serious underlying problems in FRJ's civil–military relations.

One of Milošević's key civil–military legacies was the continued role of the VSO as the supreme body in command of the Army. This council comprised of the federal, Serbian and Montenegrin presidents who in theory exercised joint command over the military.²² This arrangement also meant that Koštunica had to use the VSO mechanism if wanted to sack Pavković. His first attempt to do so in March 2002 was thwarted when it came up against resistance from his fellow VSO members Milan Milutinović and Milo Đukanović. The council met again on 24 June and again refused to support the president unless he also removed other key personnel from the VJ leadership, including the chief of the military security services, and then Koštunica ally, General Aco Tomić. Koštunica rejected this proposal and attempted to bypass the VSO and use a constitutionally ambiguous presidential decree to relieve Pavković of his duties and replace him with his deputy, General Branko Krga. Pavković declared the decision illegal and refused to go, but was left isolated after Koštunica obtained expressions of support from key generals. Pavković took his case to the constitutional court, but it was dismissed on 11 July 2002 on the grounds that the court was not competent to rule in such matters.²³

The events surrounding Pavković's dismissal highlighted three key problems in SCG's civil–military relations. First, they illustrated important problems and uncertainties in the VSO arrangement. These resulted both from the deliberately imprecise institutional legacy of the Milošević period, and also from continued uncertainty and suspicion over the balance of power between the different political entities that made up the FRJ. Second, they showed the way in which the effective operation of FRJ's mechanisms for civilian control over the military could be held hostage to domestic political disputes. Koštunica's opponents in the DOS had been calling for Pavković's dismissal for some time, but were prepared to jeopardise this through Đukanović and Milutinović's refusal to support him in the VSO. This strategy had the short-term political advantage of damaging Koštunica, but in doing so it threatened to undermine the FRJ's mechanisms for civilian control of the military.²⁴ Finally, Pavković's apparent willingness to defy the orders of the president suggest that, until June 2002 at least, civilian control over the VJ had not been properly consolidated.

Despite the chaos and political bitterness surrounding the affair, however, Pavković's dismissal had ultimately positive implications for the democratisation of FRY's civil–military relations. It demonstrated that in the final analysis the VJ was willing to submit to the principle of civilian control. Despite Pavković's considerable efforts to protect his position, the Army's leadership united behind the president and publicly declared their 'full readiness to ... implement all decisions taken by the Supreme Defence Council and the federal president'.²⁵ The removal of Pavković also removed a persistent and often vocal irritant from the SCG civil–military relationship. Pavković's replacement, General Branko Krga, has been reluctant to comment publicly on questions of domestic policy beyond those which fall under his specific remit as Chief of General Staff.²⁶ Subsequent events have also suggested that civilian control over the VSCG is a reality. Indeed, in the wake of the assassination of Prime

Minister Đinđić by a semi-autonomous police special forces unit in March 2003, the VSO proved itself able to implement further sweeping personnel changes in the Army's leadership, dismissing included 16 high ranking military officers including the controversial General Tomić.[27]

Democratising Civil-Military Relations

The removal of Pavković was an important first step towards consolidating civilian control over the military in FRJ. However, democratic civil–military relations involve more than the simple maximisation of civilian power over the Army and its non-interference in domestic politics. They also entail effective democratic governance of the defence sector as a whole. This includes a clear framework of constitutional responsibilities; mechanisms for the effective, transparent and accountable implementation of defence policy and the defence budget; and the wider engagement of civil-society in defence matters.[28] Since October 2000, FRJ/SCG has made some progress towards implementing democratic control over its armed forces. However, three factors have hampered its effective consolidation: first, continuing uncertainty and divisions over the future of the State Union;[29] second, a lasting period of political turmoil following the assassination of Đinđić in March 2003; finally, weaknesses in institutional capacity and willingness to actually implement reform in practice across both civil and particularly military sectors.

Uncertainties over the future constitutional shape of FRJ had a clear impact on the pace of civil–military reform in the between 2000–3. While this remained unclear, policy makers were reluctant to adopt new federal laws that might prove only to be temporary. In this period, for example, the federal parliament passed only one much criticised piece of civil–military legislation— the Law on the Security Services of the FRJ.[30] The creation of the State Union in February 2003 offered hope that the new constitutional arrangement would provide a firmer basis for legislative and institutional reform. This initial optimism appeared to be confirmed when Boris Tadić—a capable reformist from the DS—was appointed federal Minister of Defence and announced a ten point plan for defence reform.[31] However, since this period, further reform has largely been stalled because of continued political divisions and instability. Đinđić's murder in March 2003 sparked a struggle for power within the DS that distracted Tadić from his ministerial responsibilities.[32] It also marked the beginning of a decline in the influence of the DS who, in February 2004, were ousted from government by a weak and divided coalition led by Koštunica and the DSS.[33] This political instability prevented the government from pursuing a firm line on any defence reform issue, with one DS insider admitting that it was 'impossible to do anything' in the context of the protracted 'election dynamic'.[34]

The precarious future of the State Union has also had a negative impact on SCG's ability to introduce a clear legislative framework for democratic civil–military relations. The VSCG's uniquely prominent position as the most significant Union level institution in the SCG has caused particular problems, meaning that any discussion of Army's future involves making a clear assumption about the future status of the Union itself. In addition, the Montenegrin government remains suspicious of the VSCG as an institution. During the Milošević period, the VJ was used to intimidate the Montenegro from making any further moves towards independence from the FRJ[35] and residual fears persist that the VSCG could be used in the same way—particularly as the Army remains largely funded by Serbia and manned by Serbs.[36] The deeply contentious nature of this issue has meant that civil–military reform issues have often quickly become the subject of heated and bitter political discussion. This in turn has hampered or prevented the implementation of effective legislation on civil–military reform issues.

The deliberately imprecise framework for civil–military relations provided by the 2003 constitutional Charter is an example of these problems. In particular, the role of the Army itself is vague and avoids any direct reference to defending the sovereignty or territorial integrity of the State Union. This is because—for some Montenegrins at least—including such a clause would have provided a potential constitutional justification for the Army to prevent any potential Montenegrin secession attempt by force.[37] This ambiguity satisfies Montenegrin worries. However, it does little to elucidate the military's future role and purpose and provides an uncertain foundation for further military reform. Similarly, the Charter redefined the VSO mechanism so that all decisions relating to command of the military must be taken on the basis of consensus between its three members.[38] This provision prevents the abuse of the VSCG by the federal president and ensures that no decision relating to the military can be made without the consent of the Montenegrin president. However, it raises serious questions about the effectiveness of the military chain of command—particularly in times of potential crisis. As Mirolsav Hadžić has observed, the 2003 constitutional Charter is 'an incomplete, inconsistent document' whose 'implementation... will directly depend on the political will of power holders in Serbia and Montenegro'.[39]

In practice such 'political will' has simply not been forthcoming, with the result that the introduction of much needed new civil–military legislation to clarify the gaps in the constitutional Charter has been stalled. Perhaps most significantly, SCG had, by October 2004, still failed to adopt either a National Security Strategy or National Defence Strategy to provide a clear basis for future reform. The question of the National Security Strategy has been particularly contentious given that the term 'security' implies a wider remit than simply external defence. The state governments of both Serbia and Montenegro argue that this goes beyond the constitutional remit of the Union's powers as it would also entail discussions of the role and purpose of the police and other internal security agencies—organisations that currently fall under the authority of the state level governments. This is a particularly sensitive issue in Montenegro where the police forces are militarised and in many respects function as the states' own independent armed forces.[40] Similar Montenegrin sensitivities also prevented the adoption of a National Defence Strategy by the National Assembly in September 2004, despite its centrality to the future of further civil–military reform and the high hopes of the Ministry of Defence.[41]

This absence of a sound institutional framework for civil–military reform— together with the politically divisive political atmosphere—has had negative implications for the development of working mechanisms for effective and accountable implementation of defence policy. In particular, it has prevented the creation of workable framework legislation within which these processes can take place. This problem has been particularly apparent in relation to the practice of parliamentary control of defence policy. Though the 2003 constitutional Charter does give the National Assembly of the State Union some important powers in this area—such as the authority to develop the defence strategy—these have in reality been limited. Perhaps, most significantly, the National Assembly's power to approve the Union defence budget is constrained by the fact over 90% of this is derived from the Serbian state and so is in practice determined by the Serbian finance ministry.[42]

Even in the context of their existing powers, however, parliamentarians' abilities to adequately scrutinise defence policy have been restricted by political divisions and a general lack of defence expertise and administrative support.[43] These problems have been compounded by a continuing lack of transparency in the defence sector. Building on a tradition of institutional autonomy and secrecy that dates back to the JNA, the VJ/VSCG has often been reluctant to release detailed information on its activities. In December 2002, for example, the then federal parliament of the FRJ passed the defence budget on the basis of a two-page document provided by the Army and delivered to them late. This was despite the fact that at that time the defence budget represented fully two thirds of the entire federal budget.[44] Finally, the persistence of deep political cleavages have also contributed to parliamentary inaction. The National Assembly failed to adopt the new Defence Strategy in September 2004 largely because of continuing disagreements between Serbian and Montenegrin parliamentarians over the nature of the State Union itself.

Similar issues of legislative ambiguity, institutional incapacity and political division have hampered the effectiveness of the Ministry of Defence. While the early part of Tadić's tenure as Defence Minister saw a number of important initiatives introduced—including the formal subordination of the Army General Staff and military security services to the Ministry of Defence—his preoccupation with wider political developments in the latter half of 2003 encouraged inertia in the Ministry. While new Defence Minister Prvoslav Davinić has attempted to reinvigorate the reform process, a number of problems still remain. Progress includes reorganising the Ministry of Defence to create fewer administrative departments, producing new draft defence legislation and introducing a reform fund initiative to help pay for military reform.[45] However, the political nature of top appointments in the ministry has led to a change of personnel throughout the its structure as Tadić-era appointees have found themselves replaced by Davinić-era ones. This has undermined expertise levels as new staff have replaced more experienced existing ones.[46]

Expertise issues have also caused tensions between military and civilian personnel. Under Milošević the Ministry of Defence had been an almost wholly military institution. Since 2000, efforts have been made towards civilianisation but these have caused their own problems. In particular, the Army has been skeptical about replacing experienced military personnel with inexperienced (often young) civilians.[47] Further problems are apparent in the evolving balance of power between the Army General Staff and the Ministry of Defence. While the Army is at least in theory now under the control of the Ministry—whereas under Milošević the reverse had in practice been the case—what this actually means in reality is less clear and the VSCG retains a significant degree of institutional autonomy. For example, the constitutional Charter stipulates that the Minister of Defence 'directs' the work of the Army whilst the VSO 'commands' it.[48] However, as Hadžić has observed this leaves the relationship between the VSO, the Minister and the General Staff unclear and creates a number of potential 'loopholes' in the chain of command.[49]

During the Milošević period, civil-society involvement in security issues was limited and the defence sector was largely closed to external scrutiny. Since October 2000, this situation has changed significantly. In particular, there are several defence, security and foreign policy related non-governmental organisations (NGO) that are very active in Belgrade: the well respected Centre for Civil–military Relations (CCMR) provides expert research, analysis and comment on all aspects of the security sector reform process; the Atlantic Council of Serbia and Montenegro has developed a close advisory relationship with Minister Davinić; while the G17 Institute runs a School for Security Sector Reform aimed at developing expertise across the SCG security sector.[50]

Nevertheless, three important problems remain which limit the depth and impact of NGO influence on defence matters. First, and as noted above, the military itself remains deeply suspicious of outside involvement in its affairs— an attitude that has constrained the extent to which even the best connected NGOs can actually influence policy.[51] Second, the divided nature of Serbian political scene has left the NGO sector itself vulnerable to politicisation. At best this can lead to questions over organisations' 'non-governmental' status. At worst it can lead to accusations of corruption and clientism.[52] Finally, and despite the expansion of this sector since 2000, defence NGOs remain limited to a small group of organisations and personalities in Belgrade itself. Often they are dependent on external (i.e. international) funding for their survival and the extent of their influence beyond a rather small expert circle is open to question. More widely, the critical engagement of the general public and media in defence matters in SCG is low. Research by CCMR has found that while defence matters are covered quite regularly by print and television media, this tends to be limited to generalised reporting of government announcements rather than in-depth analysis or discussion of the issues at hand.[53]

Military Reform

The experience of war and authoritarianism had a major impact on the organisational development of the VJ in the 1990. It was isolated from the common military reform agendas of the rest of postcommunist Europe and there was no real attempt at military modernisation during this period. As a consequence, in October 2000, the VJ faced a number of pressing reform challenges. These included the redefinition of its role and purpose in SCG's new security environment, re-equipping and restructuring in line with this and the challenge of professionalisation. In practice, progress in these areas has been quite limited. The uncertain future of the State Union and the failure to finalise key defence legislation has made further discussion of the VSCG's future roles and missions problematic. Moreover, the period since October 2000 has seen sweeping cuts in the defence budget and the defence reform process as a whole faces severe financial constraints. These pressures have led to an ad hoc and reactive approach to military reform issues that in practice have left them in the hands of the military institution itself. Despite these obstacles, however, an identifiable direction for the VSCG's military reform process *has* emerged even if to date this has not translated very far into concrete policy. This broadly follows the path of other central and eastern European military reform programmes and envisages a smaller, more mobile, and more professional VSCG.

In October 2000, the VJ was by common agreement oversized and much of its equipment obsolescent or war damaged.[54] The fall of Milošević created the political space for the Army's leadership to begin to think about the need for reform, and in October and December 2001, the VSO announced a series of programmes aimed at restructuring the VJ. These were developed wholly within the military itself with no consultation with either the civilian sector or any external experts. They included a plan to reorganise the VJ's force structure, reduce its manpower strength to around 60,000 by 2005, and shorten the conscription period from 12 to 9 months.[55] Despite these apparently positive moves however, many critics at the time suggested that that these downsizing and restructuring plans were largely superficial and simply removed those parts of the Army that had already atrophied. Similarly, they argued that the decision to reduce the length of the conscription period was taken for short term economic reasons and not tied to any broader vision of the practical requirements of the Army's roles. Indeed, it is significant that soon after coming into office in 2002, Chief of Staff Krga himself stressed that for long term military reform to be possible the Army must wait for the approval of a

National Security Strategy that would in turn lead to a coherent Defence Doctrine and Reform Plan.[56]

The early period of Tadić's tenure as Minister of Defence in 2003 indicated a positive shift in the government's approach to military reform. Crucially, this appeared to be both proactive and civilian driven. Tadić's vision was for a slimmed down VSCG, with a focus on professionalisation as 'a precondition for creating interoperable armed forces that can undertake humanitarian and peacekeeping activities'.[57] Tadić's plan placed the defence reform process specifically within the context of the potential membership of SCG in PfP and prioritised the finalisation of key defence legislation as a first step in this process.

Certainly, the need for reform was increasingly pressing as conditions within the military deteriorated in line with falling the defence budget. Between 1996 and 2000, for example, defence spending had been sustained at over 8% of GDP. In contrast, the defence budget for 2001 was 4.6% of GDP, falling to 3.3% in 2004.[58] The impact of these spending cuts on the military itself has been severe. In September 2002, for example, the VJ claimed that it had already spent 98% of its budget for the 2002 fiscal year, and announced that it would have to send around a quarter of its soldiers home every weekend to save on food rations.[59] Moreover, almost 80% of the defence budget in 2004 was allocated towards personnel and maintenance costs, leaving only 20% for all other expenses including procurement and research and development.[60] These problems are exacerbated by extremely poor housing conditions, salaries and living standards of servicemen and—with an average age of 41—a rapidly aging demographic profile and top heavy rank structure.[61] These difficulties have sharpened the need for rapid restructuring so that the Army can live within its means. However, they also place serious financial constraints on any modernisation or reform programme that attempts to address these issues.

The political upheaval and uncertainty brought about by Ðinđić's murder and the subsequent struggle for power in Belgrade made clear civilian leadership on the military reform issue difficult and hindered the implementation of Tadić's plans. This was particularly the case as far as the introduction of new defence legislation was concerned. Nevertheless, Davinić has committed himself to the spirit of Tadić's ten point programme and the contours of a future reform strategy have emerged. This envisages a three stage plan for reform that will take the Army to 2010. It will focus on increasing the proportion of professional, all volunteer personnel in the VSCG, withdrawing obsolescent equipment and selectively replacing it with new materiel, and reducing the overall size of the VSCG to between 40,000 and 50,000 strong.[62]

Downsizing in particular has emerged as a key reform strategy because it appears to offer at least part of a solution to the VSCG's financial woes. In cutting back on the manpower of the Army, the hope is that resources will be freed up for modernisation while at the same time providing an opportunity to further professionalise the remaining personnel. The government has introduced a 'Reform Fund' in an effort to help finance these reforms. This will be drawn from money raised by the sale of some of the Army's extensive real estate holdings and will be targeted specifically at addressing the military's housing crisis and financing the resettlement costs of the downsizing programme.[63] There also seems to be broad consensus on the future roles of the VSCG—even if for the moment this question is complicated by the unpredictable future of the State Union. These focus around three main missions. First, defence of national territory from external threat; second, the ability to contribute to multinational peacekeeping operations; and, finally, the provision of support to other government departments in addressing internal security threats.[64]

There remain a number of serious concerns about the VJ's current programme of military reform, however. By their own admission, the absence of framework legislation has forced the Army into a 'tactical' rather than 'strategic' approach to the issue.[65] While this is perhaps understandable given the current domestic political context—and many argue that some reform is better than none at all—it has led to a piecemeal, reactive approach to change rather than a planned, holistic process of reform. It also leaves the process almost wholly in the hands of the Army itself—something many analysts identify as a real problem given its association with the old regime and its history of 'outdated thinking'.[66] This situation is unlikely to change until the civil sector can introduce appropriate defence legislation to provide an adequate basis for a civilian-led reform process. At the time of writing (March 2005) this process, has begun though key documentation such as the National Security Strategy and the Military Doctrine remain stalled.[67]

Even with suitable legislation in place, military reform is unlikely to be either easy or straightforward. As the Croatian experience discussed elsewhere in this volume illustrates, the socio-economic costs of 'downsizing' thousands of soldiers are likely to be high, while professionalised volunteer soldiers cost much more than conscripts both to train and pay.[68] The proposed Reform Fund might go some way towards meeting these costs, but worries remain over the potential implications of selling off so much of the Army's—admittedly bloated—infrastructural capacity without proper planning, and the possible vulnerability of the Fund to embezzlement and corruption.[69] Any professionalisation process will also require deep and difficult attitudinal change in the Army itself. In contrast to many other communist-era armed forces, and in relation to expertise at least, the JNA maintained a relatively strong professional ethos that continues to this day in the VSCG. It also has a well-established military education system that—while somewhat dated and conservative—has the potential to form the basis of any future professionalisation process. Nevertheless, as James Gow has observed, the Army's failure to properly acknowledge its negative role in the Yugoslav conflict—and particularly its link to war crimes—continues to undermine those elements of professionalisation that embrace ethics and responsibility.[70]

The VSCG's new missions present new opportunities, but also a number of potential difficulties. Peacekeeping, for example, could offer the Army some important advantages. It would build on the old JNA's positive traditions in this area; help to 'normalise' external perceptions of the post-war Army; and could act as an important foreign policy tool for building relations with NATO and the US. However, deploying and sustaining significant numbers of troops abroad is expensive and requires capabilities—such as strategic airlift—that the VSCG does not currently possess. SCG's still troubled relationship with the ICTY is also a significant obstacle and many potential peacekeeping partners have been wary of cooperating too closely with a military that remains associated with war crimes. The VSCG's contribution to multinational peacekeeping operations since 2000 has therefore been limited to a small number of military observers and medical personnel. Proposals to contribute larger numbers of troops to the peacekeeping missions in Liberia and Afghanistan have in practice foundered in the face of political division, resource constraints and international scepticism.[71] Similarly, the development of the VSCG's internal security role has occurred partly in response to the very real security threat posed by ethnic Albanian secessionist groups in South Serbia and—in the context of responses to 9/11—is reflective of defence reform trends elsewhere in Europe and the world. However, this mission remains potentially problematic, given the JNA/VJ's chequered history in this area during the Yugoslav conflict, current debates over the future status of Montenegro in the State Union, and the already crowded nature of SCG's security sector.

Finally, the extent to which the Army itself is committed to reform remains an open question. Publicly at least, Krga and the general staff are supportive of the process, recognising that military professionalisation and modernisation offer a path out of the isolation and humiliation of the 1990s. Likewise, for a younger generation of officers, professionalisation and membership of PfP offer the hope of a brighter future and a more fulfilling career.[72] Nevertheless, any serious reforms will inevitably threaten the jobs and livelihoods of many thousands of military personnel—particularly in the top heavy middle ranks—and passive resistance to change from these cadres is likely.[73] They will also threaten the Army's institutional autonomy and tradition of exclusive control over its own internal affairs. How far these challenges will actually impact on the success of any reform process is uncertain. At the very least they indicate some of the serious obstacles reformers will have to overcome if their plans are to be successful and suggest that implementing change in practice is likely to be a long-term and difficult process.

The International Community

The domestic political context has been of key importance in dictating the pace and extent of SCG's civil–military reforms. Nevertheless, this reform process has also taken place in the context of intense international interest and political pressure. The SCG government has publicly identified integration with western institutions—and particularly membership of NATO's PfP programme—as a foreign policy goal. In turn the international community has been eager to try and encourage democratisation and regional stabilisation in the whole of the former Yugoslav region through assistance programmes and political conditionality. Defence reform issues have been central in all these areas, and this factor has placed SCG's civil–military relationships under unprecedented international scrutiny. Nevertheless, SCG's relations with the international community have not always been easy, with the issue of cooperation of with the ICTY proving particularly divisive.

The legacy of the 1990s—in particular of sanctions and the NATO bombing of 1999—has contributed to a continuing reticence towards western institutions in SCG. Indeed, it was only in April 2002 that the government finally announced its intention to apply to join PfP. This was in contrast to most other states in the former Yugoslav region for whom PfP membership had been a key foreign policy priority for some time.[74] Despite this initial reticence, western integration for SCG has become an increasingly important element of the government's foreign and defence policy since the beginning of 2002. This has been particularly the case in relation to SCG's potential membership of the PfP, though it has also played an important role in its application to join the Council of Europe and its attempts to negotiate an Stabilisation and Association Agreement with the EU.[75] The pre-conditionality requirements of each of these institutions have at their heart issues closely linked to defence reform. Broadly, these concern the consolidation of civilian control over the Army, and especially full cooperation with the ICTY.

NATO has been clear that there are several non-negotiable steps that SCG must take if it is serious about joining the PfP programme. These have included or continue to include: an end to the Army's links with the Bosnian Serb Army (VRS), including financial support; support for all aspects of the Dayton agreement, including ratification; personnel change within the Army to remove figures closely associated with the Milošević regime or implicated in war crimes; the withdrawal of the government's lawsuit against NATO with the International Court of Justice over its bombing campaign of 1999; full cooperation with the ICTY; and full compliance with UN Security Council Resolution 1244 on Kosovo.[76] Broadly, these also reflect the demands of the Council of Europe and the EU in the defence sphere. Since 2000, the government has taken steps in all of these areas. For example, at the beginning of March 2002 Belgrade ended all official financial aid to the Ministry of Defence of the Republika Srpska and the VRS, while in December the federal parliament ratified the Dayton peace accords.[77] Moreover, a realisation that Pavković's continuation as Chief of General Staff would always be an insurmountable obstacle to PfP membership appears to have been a key contributory factor in Koštunica's decision to instigate his removal between March and July 2002. Koštunica's first unsuccessful attempt to dismiss Pavković occurred during the 25 March meeting of the VSO, when membership of the PfP was first officially suggested.[78]

Nevertheless, whatever the attractions of potential membership of PfP and other western institutions, and whatever progress has been made in the pre-conditionality demands outlined above, the issue of cooperation with the ICTY—and particularly the arrest and the arrest and transfer of former VRS Chief of General Staff and ICTY indictee General Ratko Mladić—remains unresolved.[79] The pre-conditionality process alone has not been sufficiently persuasive to encourage compliance in this area. The war crimes legacy and the question of cooperation with the ICTY goes far beyond the confines of the VJ/VSCG and the defence reform process. It incorporates elements from across the political and security sectors of both SCG and the Republika Srpska, including politicians, the police and the VRS. Nonetheless, the issue of the Tribunal remains of particular relevance to SCG's defence reforms. The Army itself was implicated in war crimes in the 1990s, and a number of former VJ personnel have been indicted by The Hague.[80]

The significance which the international community has placed on SCG's cooperation with the ICTY is illustrated by the extent to which direct conditionality has been used to encourage compliance. This has taken the form of threats to withdraw incentives or inflict punitive measures should either state not implement particular policies or obligations. Direct conditionality differs from pre-conditionality in that it is not a process which the target state opts into, but one that is imposed upon it. This policy played a crucial role in the attempts of the US and other states to put pressure on the SCG authorities to arrest Slobodan Milošević in March 2001, and then to subsequently transfer him to The Hague.[81] Direct conditionality also played an important role in pushing the government to introduce new legislation on cooperation with the ICTY and issuing domestic arrest warrants for war crimes suspects in 2002.[82] Most recently, in Autumn 2004, the United States threatened to cut off financial aid to Serbia–Montenegro and block its membership of international financial institutions because of its belief that Belgrade is not moving far or fast enough in relation to its cooperation with the Hague. These threats coincided with an apparently reinvigorated approach to the ICTY issue in SCG. Key initiatives included the arrest of Ljubiša Beara—a former VRS officer implicated in the Srebrenica massacre of 1995—and an attempt by Serbian anti-terrorist police to apprehend Mladić himself.[83]

However, beyond forcing the Belgrade authorities to take explicitly mandated steps—such as the arrest of specific ICTY indictees—the impact of direct conditionality on the civil–military reform process more widely has been limited. Change in this area is dependent on far wider societal and political attitudinal change in Serbia as well as within the military itself. In particular, the war legacy remains deeply contentious. There has not yet been any broad public or political recognition of the responsibility that the Serbs and Serbia may hold for the war itself—and certainly not for many of the atrocities perpetrated during its course. As a consequence, perceptions of the war remain coloured by the nationalism of the 1990s. The ICTY remains deeply unpopular, and there is a widespread perception that the whole process is discriminatory and anti-Serb.[84] For its part, the Army remains deeply resistant to any implication of its involvement in war crimes and highly suspicious of the whole Hague process. This underlying context has had important implications for willingness and ability of civilian politicians to push through reform in these sensitive areas, particularly given the weakness and hence political vulnerability of each of the post-Milošević coalition governments. Many fear that pushing too hard on the ICTY issue risks handing support to radical nationalist parities and undermining the reform agenda more widely.[85] For this situation to change, there will need to be a much longer-term attitudinal shift at all levels of Serbian society, and this is something that is unlikely to be imposed from above by the use of conditionality.

Conclusion

Since the fall of Milošević, SCG has made important progress in key areas of civil–military reform. The principle of civilian control over the armed forces and their non-interference in domestic politics has been established. A broad vision for the future direction of military reform and defence policy has emerged that envisages a smaller, more professional VSCG, able to participate in multinational operations and operating in the context of PfP membership. Some reorganisation of the defence sector has taken place and important defence legislation has been drafted. These steps have provided an underlying basis for the establishment of democratic control over the armed forces in SCG and broadly reflect military reform trends visible elsewhere in postcommunist Europe.

Nevertheless, successive governments in Belgrade have struggled to build on these foundations and civil–military reform efforts remain unconsolidated. In particular, domestic political divisions—both within the DOS and the subsequent, Koštunica-led governing coalition and between Serbian and Montenegrin politicians over the future status of the State Union—have prevented the finalisation of a coherent civilian-led defence reform programme strong enough to challenge the military's own traditions of institutional autonomy. In this policy vacuum, further civil–military reforms have been difficult. When they have taken place they have tended to be initiated by the military itself and occurred in a conservative, ad hoc and fragmented manner. In practice, the military remains secretive, oversized and outdated. The Ministry of Defence is inefficient and lacks administrative capacity and civilian expertise. The SCG parliament's ability to provide oversight and scrutiny of defence policy is hampered by similar problems, while civil society engagement in defence policy issues is limited to a small number of organisations and constrained by a lack expert media analysis. Integration with western institutions and particularly membership of PfP remain possibilities, but will not happen until the government proves itself more cooperative on the ICTY issue.

The evolution of civil–military relations in SCG have been conditioned the wider domestic context. This in turn has been shaped both by the persistent legacies of the Milošević period and by the continuing failure to resolve the question of legitimate statehood in the wake of the collapse of the SFRJ. This wider context has helped to push military reform to the back of the political agenda and left it in the hands of the conservative and secretive military institution itself. The result has been stagnation. While the international community has tried to support and encourage the reform process, in practice it has been unable to force it. SCG's experiences in this area illustrate that while defence reform may be an important part of wider democratisation and post-conflict transformation processes, it is also to a very large extent dependent on them for its own success.

Notes

[1] The research for this paper was carried out with the kind financial support of the Geneva Centre for Democratic Control of Armed Forces and the British Academy Small Research Grants Scheme. This paper draws in part on Timothy Edmunds, *Defence Reform in Croatia and Serbia–Montenegro* (Oxford: Oxford University Press, 2003).

[2] The FRJ was succeeded on 4 February 2003 by the State Union of Serbia and Montenegro (henceforth SCG). This is a unique constitutional arrangement under which power is substantially devolved to each of its two member states of Serbia and Montenegro. The State Union—which has its own president, ministries and parliament (the National Assembly)—is the sovereign entity in the international arena but in practice is only responsible for foreign and defence policy. To complicate matters further, Kosovo still technically remains a province of Serbia though in practice is run by the United Nations Interim Administration (UNMIK). See *Constitutional Charter of the State Union of Serbia and Montenegro*, 4 February 2004. Available at http://www.mfa.gov.yu/Facts/const_scg.pdf (accessed October 2004).

[3] The Army in SCG has been through several name changes in since the early 1990s. In 1992, the old Yugoslav People's Army (JNA) was disbanded and split into the Bosnian Serb Army of Republika Srbska (VRS) and the Yugoslav Army (VJ) of the FRJ. This division occurred in an attempt to distance Belgrade from Bosnian Serb activities in Bosnia and sidestep international sanctions. In February 2003, with the creation of the State Union of Serbia and Montenegro, the VJ became the Armed Forces of Serbia and Montenegro (VSCG).

[4] Timothy Edmunds, *Defence Reform in Croatia and Serbia–Montenegro*, Adelphi Paper 360 (Oxford: Oxford University Press, 2003), pp. 5–8.

[5] James Gow, *Legitimacy and the Military: The Yugoslav Crisis* (London, Pinter Publishers, 1992), p. 59. Other good studies of the history of the JNA and its role in the war years include: James Gow, *The Serbian Project and its Adversaries: A Strategy of War Crimes* (London: Hurst & Company, 2003); Miroslav Hadžić, *The Yugoslav People's Agony: The Role of the Yugoslav People's Army* (Aldershot: Ashgate, 2002); Biljana Vankovska & Håkan Wiberg, *Between Past and Future: Civil–Military Relations in the Post-Communist Balkans* (London: I.B. Tauris, 2003).

[6] James Gow, 'Professionalisation and the Yugoslav Army', in Anthony Forster, Timothy Edmunds & Andrew Cottey', *The Challenge of Military Reform in Postcommunist Europe: Building Professional Armed Forces* (Houndmills: Palgrave-Macmillan, 2002), pp. 183–185.

[7] Article 135, *Constitution of the Federal Republic of Yugoslavia*, 1992. See also Dimitrios Koukordinos, 'Constitutional Law and the External Limits of the Legal Framing of DCAF: The Case of Croatia and the FR Yugoslavia', in Biljana Vankovska (ed.), *Legal Framing of the Democratic Control of Armed Forces and the Security Sector: Norms and Reality/ies* (Belgrade: Centre for Civil–military Relations and the Geneva Centre for the Democratic Control of Armed Forces, 2001), pp. 164–165.

[8] James Gow, 'The European Exception: Civil–military Relations in the Federal Republic of Yugoslavia', in Andrew Cottey, Timothy Edmunds & Anthony Forster (eds), *Democratic Control of the Military in Postcommunist Europe: Guarding the Guards* (Houndmills: Palgrave, 2002), pp.57–58. See also, Robin Alison Remington, 'The Yugoslav Army: Trauma and Transition', in Danopoulos & Zirker (eds), *Civil–Military Relations in the Soviet and Yugoslav Successor States*, p. 167; Gow, 'The European Exception', pp. 201–220.

[9] Remington, 'The Yugoslav Army', p. 167; Gow, 'The European Exception', pp. 202–203.

[10] Gow, 'The European Exception', pp. 205–206.

[11] Ibid., pp. 209–10; Budimir Babović, 'Police as a Tool of Milošević's Autocratic Rule', *In the Triangle of State Power: Army, Police, Paramilitary Units* (Belgrade: Helsinki Committee, 2001), pp. 24–31. Available at http://www.helsinki.org.yu/doc/pubs/files/eng/HFiles9.zip.

[12] 'Pavkovic: Yugoslav Army Ready to Defend Country', *RFE/RL*, 22 September 2000.

[13] Koštunica was federal president until March 2003 and is currently Serbian Prime Minister. Đinđić was Serbian Prime Minister until his assassination in March 2003. For more on the DOS see International Crisis Group, *Serbia's Transition: Reforms under Siege*, ICG Balkans Report No. 117, 21 September 2001. Available at http://www.icg.org.

[14] Interviews, Belgrade, June 2002, June 2004. See also Miroslav Hadžić, 'Civil–military Features of the FR Yugoslvia', *CCMR Research Paper*, 23 August 2002,p.1. Available at http://www.ccmr-bg.org.

[15] *Human Rights in Transition: Serbia 2001*, Helsinki Committee for Human Rights in Serbia Annual Report 2001 (Belgrade: Helsinki Committee, 2002). Available at http://www.helsinki.org.yu/report.php?lang=en (accessed); Hadžić, 'Civil–military Features', p.2.

[16] Interview with Ivan Vejvoda, Executive Director, Balkan Trust for Democracy, 22 June 2004.

[17] Pavković was formally indicted by the ICTY in October 2003.

[18] 'Yugoslav Defence Minister Quits', *RFE/RL*, 16 January 2002; 'Yugoslav Military Contradicts Foreign Minister, *RFE/RL*, 19 February 2002.

[19] *Serbia: Military Intervention Threatens Democratic Reform*, ICG Balkans Briefing, 28 March 2002, pp. 2–3; Edmunds, *Defence Reform*, pp.29–30; 86 fn. 63.

[20] For more on the Koštunica/Pavković relationship see Hadžić, 'Civil–military Features', p.5.

[21] For more details see Timothy Edmunds, 'Crisis or Turning Point? Koštunica, Pavković and Civil Military Relations in FRJ', *CCMR Research Paper*, June 2002. Available at http://www.ccmr-bg.org.

[22] Ultimate command responsibility in the VSO remained deliberately ambiguous, however, with the federal president commanding the Army 'in line with decisions from the VSO'. See Miroslav Hadžić, 'New Constitutional Position of the Army', *DCAF Working Paper No. 112* (Geneva: Geneva Centre for the Democratic Control of Armed Forces, February 2003), pp. 8–11. For further discussion of the VSO, see Edmunds, *Defence Reform*, pp. 26–27.

[23] *B92 News Archive*, 24, 25, 26 June 2002; 11, 12 July 2002. Available at http://www.b92.net.

[24] At the time, the DS actually went so fact as to accuse Koštunica of abusing the VJ. See Hadžic, 'Civil–military Features', p. 3.

[25] 'Army Officers Abandon Sacked Leader', *B92 News Archive*, 26 June 2002.

[26] Krga has not remained completely silent, however. See, 'Head of Serbia and Montenegro's Army Proposes Return of Armed Forces to Kosova', *RFE/RL*, 22 August 2003. 'Serbia and Montenegro's Defence Minister Doubts Return of Troops to Kosova', *RFE/RL.*, 26 August 2003.

[27] 'Supreme Defence Council Dismisses Army Intelligence Chief', *B92 News Archive*, 21 March 2003; 'Serbia Montenegro Sacks 16 Old Guard Officers', *RFE/RL*, 8 August 2003.

[28] Andrew Cottey, Timothy Edmunds & Anthony Forster, 'The Second Generation Problematic: Rethinking Democracy and Civil–military Relations', *Armed Forces and Society*, 29: 1 (Fall 2002).
[29] SCG's current constitutional arrangement was agreed only after strong pressure from the EU. Montenegro retains a large constituency in favour of independence and will be able to vote to leave the Union in 2006. See Filip Vukanović, 'EU Realising State Union Can't Survive', *Institute for War and Peace Reporting—Balkan Crisis Report*, 21 September 2001.
[30] See *The Law on the Security Services of the Federal Republic of Yugoslavia*, 3 July 2002. Available at http://www.osce.org/documents/FRJ/2002/07/129_en.pdf.
[31] Boris Tadić, 'Reform of the Defense System of Serbia and Montenegro', *Statement by the Minister of Defence at the Assembly of Serbia and Montenegro*, Belgrade, 21 March 2003.
[32] Interviews, Belgrade, June 2003. Tadić was ultimately elected Union President in June 2004.
[33] The coalition is made up of five parties including moderate nationalists, reformists, and radical socialists from Milošević's old party.
[34] Interview with Vejvoda.
[35] Gow, 'Professionalisation', pp. 189–190.
[36] Serbia will provide around 95% of the Union defence budget in 2004. Amadeo Watkins, 'PfP Integration: Croatia and Serbia and Montenegro', *CSRC Research Paper 04/05* (Camberley: CSRC, April 2004), p. 15.
[37] Article 53, *Constitutional Charter*.
[38] Article 56, *Constitutional Charter*.
[39] Hadžić, 'New Constitutional Position', p. 29.
[40] Interview with Pavle Janković, Assistant Minister, Ministry of Defence of SCG, 23 June 2004.
[41] 'No Support for Defence Plan', *Beta News Agency*, 21 September 2004. The Defence Strategy was finally passed on 18 November 2004 as this paper was going to press.
[42] Hadžić, 'New Constitutional Position', pp.18–19.
[43] Interviews, Belgrade, June 2002; June 2004. See also Svetlana Djurdjević-Lukić, 'Defence Reform in Serbia and Montenegro: Unrealistic Expectations', unpublished MA Dissertation, SSEES, University College London, May 2004, p. 21.
[44] Svetlana Djurdjević-Lukić, 'President's Men', *Transitions Online*, 15 March 2002.
[45] Interview with Janković.
[46] Interviews, Belgrade, June 2004.
[47] Interview with Radovanović, Deputy Minister for Foreign Military Cooperation and Defence Policy, federal Ministry of Defence, 20 June 2002; interviews with senior military officers, Belgrade, June 2004.
[48] Article 41, *Constitutional Charter*.
[49] Hadžić, 'New Constitutional Position', pp. 25–26.
[50] See for, example, http://www.ccmr-bg.org; www.g17institute.com
[51] Interview with Vladan Živulović, President of the Atlantic Council of Serbia and Montenegro, 22 June 2004. Živulović characterised the VSCG as 'a state within a state'.
[52] Interviews, Belgrade, June 2004.
[53] Jocvanka Matić, 'Media Reporting on Defence and the Army', in *The Serbian and Montenegrin Public on Reform of the Army—1st Round* (Belgrade: Centre for Civil Military Relations, 2003), pp. 38–40. Available at http://www.ccmr-bg.org.
[54] The Air Force alone lost some 36% of its fighter aircraft and 60% of its air defence radar during the NATO bombing of 1999. Aleksandar Radić, 'Modernisation of the Yugoslav Army', *CCMR Research Report*, 2002.
[55] Edmunds, *Defence Reform*, pp. 47–48.
[56] Branko Krga, 'Yugoslav Army Reform: Achievements', in Miroslav Hadžić (ed.), *Armed Forces Reform—Experiences and Challenges* (Belgrade: Centre for Civil–military Relations, 2003), p. 66.
[57] 'Interview with Boris Tadić'. *Janes Defence Weekly*, 16 July 2003, p.32; Tadić, 'Reform of the Defense System'.

[58] Estimated figures. *The Military Balance, 1997–98 to 2001–02* (London: International Institute for Strategic Studies, 1997–2001). 2004 figure provided by the Ministry of Defence of Serbia and Montenegro, 23 June 2004.
[59] 'Yugoslav Army to Send Troops Home to Save Food Rations', *RFE/RL*, 6: 180 (24 September 2002).
[60] *Draft White Paper on the Defence of Serbia and Montenegro* (Belgrade: Ministry of Defence, June 2004), pp. 31–32.
[61] *Draft White Paper*, pp. 24–25; Watkins, 'PfP Integration', p. 19.
[62] Watkins, 'PfP Integration', p.17; Branko Krga, 'Reform of the Serbian and Montenegrin Army', *CCMR Analysis*, 2003; available at http://www.ccmr-bg.org.
[63] Interview with Janković.
[64] Interview with General Branko Krga, Chief of Staff of the VSCG, 23 June 2004.
[65] Interview with Krga.
[66] Interview with Miroslav Hadžić, quoted in: Zoran Kusovac, 'Belgrade's Battle for Change', *Jane's Defence Weekly*, 16 January 2002, p. 23.
[67] See for example *Draft White Paper*.
[68] See for example this author's Croatia case study (with Alex J. Bellamy) elsewhere in this special issue.
[69] Interviews, Belgrade, June 2004.
[70] Gow, 'Professionalisation', pp. 187–188.
[71] For an excellent discussion of these dilemmas see: Nick Webb, 'Involvement with Multinational Forces: An Indication of Serbia's International Standing?', unpublished MA Dissertation, Department of Politics and International Relations, University of Lancaster, 2004.
[72] Interviews with serving military officers, Belgrade, June 2004.
[73] A serving Colonel interviewed by the author in June 2004 characterised the reform process as 'total anarchy' and the reformers as 'criminals', noting that 'all we [in the Army] can do is be angry and frustrated'.
[74] 'Yugoslavia Seeks Membership of Partnership for Peace', *RFE/RL*, 26 April 2002.
[75] Serbia and Montenegro was finally admitted to the CoE on 3 April 2003; 'Council of Europe To Admit Serbia and Montenegro', *RFE/RL*, 27 March 2003.
[76] Interview with Nowosielski.
[77] 'Yugoslav Army to Cut off Military Aide to Bosnian Serb Army', *RFE/RL*, 8 February 2002; Yugoslavia Approves the Dayton Agreement', *RFE/RL*, 18 December 2002. See also, 'Belgrade Ends Military Links to Banja Luka', *RFE/RL*, 9 April 2003.
[78] 'Yugoslav Army Chief's Fate to be Determined'; 'And Yugoslav Leaders Suggest Joining NATO's Partnership for Peace', *RFE/RL*, 26 March 2002.
[79] See, for example, 'US Group Says Only One Man Stands between Serbia and NATO', *RFE/RL*, 15 July 2003.
[80] Gow, *The Serbian Project*, pp. 75–79.
[81] The decision to extradite Milošević to The Hague on 28 June 2001 occurred in the context of an aid donor's conference for FRJ on 29 June 2001. Both the US and EU had made it plain to the Yugoslav authorities that a substantial aid package amounting to $1.28 billion in grants and loans would be at risk if the extradition did not take place.
[82] See for example, 'Yugoslav Parliament Passes Controversial Hague Cooperation Law', *RFE/RL*, 12 April 2002; 'Belgrade Gives War Criminals Deadline', *RFE/RL*, 18 April 2002; 'US Lifts Freeze on Aid to Yugoslavia', *RFE/RL*, 22 May 2002.
[83] 'Prime Suspect in Srebrenica Massacre Transferred to the Hague', *RFE/RL*, 12 October 2004; 'Serbs Step up Search for Mladić before Deadline to Cut off Aid', *The Independent*, 29 September 2004.
[84] Watkins, 'PfP Integration', p. 13.
[85] It also risks their lives. Many speculate that Prime Minister Đinđić was assassinated in 2003 as a direct result of his intention to step up cooperation with the ICTY.

Vladimir Putin and Military Reform in Russia

DALE R. HERSPRING
Department of Political Science, Kansas State University, Manhattan, USA

From all appearances, President Vladimir Putin has become serious about military reform. This does not mean that we should expect radical or revolutionary policy changes in this area however. Indeed, as Lilia Shevtsova pointed out in her recent book on Putin, whereas Boris Yeltsin was a revolutionary, a man who destroyed the pre-existing communist political system, Vladimir Putin is a bureaucrat, a man who considers his primary task to bring stability to Russia.[1] The problem, however, is whether this bureaucrat will remain primarily a 'stabilizer' or whether he is prepared to move closer to reform by becoming a 'transformer; a pioneer reorganizing for the first time in history the way Russia is ruled'.[2]

Before continuing, it is necessary to say something about Putin, the politician. I argue that there are five factors that characterize Putin's approach to politics.[3] First, as noted above, Putin is a bureaucrat. Putin believes that the leader at the top should be able to set the system's parameters but, once having done so, he would prefer to leave policy implementation to the bureaucrats. However, he recognizes that the bureaucracies often work to subvert a political leader's wishes. This is especially true of the military that tends to be very conservative and focused on the past. Generals and admirals often like doing things 'the old way'. Getting them to accept new approaches generally means imitating Lenin's famous dictum of 'Two Steps Forward and One Step Backward'. Forcing a bureaucracy to change is a slow, frustrating process, but Putin believes that such organizational structures can only be changed by continually pushing them, by gradually changing the structure, attitudes and personnel in the bureaucratic system. Given its highly bureaucratized nature, Putin fully understands that it cannot be turned upside down as the Bolsheviks did the Tsarist Army during the Civil War. He has to work with what is available. This is why I believe that those observers (including this writer) who expected Putin, and his hand-picked defence minister, Sergey Ivanov to take the kind of 'bold' decisions necessary to make military reform a reality in a relatively short period of time were mistaken. Bold decisions are not part of Putin's leadership style.[4]

Putin's second characteristic is respect for Russian political culture. While he may be seen as a 'Westernizer' of the Peter the Great type by many in the West, he believes that it would be wrong to force the Russian military (or any other part of Russian society) to mimic the West. He wants to move the Russian military closer to the kind of system found in the West, but he knows that in the end it will continue to have its Russian idiosyncracies.

The third factor that characterizes Putin's leadership style is his pragmatic approach to policy. When he was a KGB officer, his primary goal was to find a way to solve problems. If a liberal idea would 'work', he would accept that. If a conservative approach worked better, he was ready to accept that approach as well. The result: he is pragmatic, and flexible when it comes to policy issues.

Fourth, Putin is *not* a long-term planner. He lives in the here and now just as he did in the KGB. This helps explain why he has not come up with a long-term plan to reform the military (or any other part of the Russian polity)—other than to push the somewhat vague concept of military professionalism. He does not conceptualize problems or answers, rather, he takes whatever the situation will permit—and that includes retreating on occasions, but always pushing for the vague idea of a professional military.

Finally, and much to the frustration of many Western observers and policy makers, Putin is cautious when it comes to making changes. His decision-making approach tends to be incremental. He knows he does not have enough power to make all the changes that Russia needs at once. The country is in too bad shape for that approach. Nevertheless, he also knows that if Russia is to survive, it must change. The state must be rebuilt if it hopes to handle the country's problems in an efficient manner. It is also worth noting that Putin will take advantage of events—as he did with 9/11—to get the military to do what he wants but, in the end, he is the opposite of a Khrushchev with his 'hair-brained' schemes. He is more the tortoise than the hare (and we all know that in the end it was the tortoise that won the race). He takes one small step at a time, always pushing the system to change, but backing off if bureaucratic resistance gets too strong.

Problems in the Russian Military

The Russian military is beset with a number of severe and long-term problems.[5] These include institutionalized *dedovshchina* or hazing practices, inadequate training; criminal behavior within the armed forces; and poor quality draftees. Hazing continues to be a major problem. Yelena Shishounova noted in January 2003, for example, that:

> Of course, hazing has not disappeared. Instead, it has worsened in many cases. The Defence Ministry steadfastly refuses to consider plans to create a professional NCO corps, but as long as there are no professional NCOs, 'grandfathers' [longer serving soldiers] will remain indispensable as an organizing force in the barracks, and officers will continue to permit hazing.[6]

In fact, the Chief Military prosecutor noted that of the list of 316 officers who had been convicted of bullying the previous years included 'one general who had beaten up a colonel'.[7]

Then there is the question of training—vital to every military. From 1992 until the year 2000, the military did not have any combat training—because of a lack of funds. Even in May 2003 for example, Russian pilots received only 6–8 hours of combat training per year, far below the 150 or so hours that NATO requires of its pilots.[8] This means that Russia now has a whole group of mid-level officers who have had little or no 'hands on' training (except for those who have served in Chechnya). The result is that 'It is not possible to make professionals of them in a couple of years.'[9] Then we have cases of senior military officers complaining about the weapons systems they have. For example, General Yuri Grishin, the Deputy Commander of the Air Force complained that the Air Force will not get new planes until 2010. In the meantime, it will have to 'fly planes that are 20 to 40 years old'.[10] Or, take the airborne forces. According to its Commander, General Georgiy Shpak, 'Outmoded BMD-1 and BTR-D vehicles, adopted into the armament in 1969 and 1974 respectively, make up the bulk of the inventory. Up to 80 per cent of them have been operated for fifteen years or more.'[11] And the situation in the Navy is just as bad. To quote the Commander of the Black Sea Fleet, Admiral Vladimir Komoyedov, 'Our Navy has been going downhill for a long while, gaining speed each year. A few years more and there may be an invisible process in about 5–7 years; we simply won't have any ships left. At least, ships that are moving.'[12] And if that were not enough, in August 2003 the Defence Ministry announced that military procurements would be cut in 2004.[13]

The military also has very serious problems both with crime and with the quality of draftees it gets twice a year. To begin with, desertion has become commonplace in many units. To quote one observer, 'Soldiers are deserting because they see no chance of a normal human existence for themselves either in or after the Army.'[14] The military claims about 5,000 desert every year.[15] The number of suicides has also increased. In 2002, for example, General Staff crime statistics suggested that 89 soldiers, warrant officers, and officers committed suicide.[16] The generals blame desertions on their commanders—who often pay little attention to what is going on in units, preferring to let longer serving soldiers exercise discipline—discipline that often results in rapes, beatings, and sometimes even death.

The situation among draftees—necessary as long as conscription exists—is not much better. The Deputy Chairman of the Defence Committee Nikolay Bezborodov complained that '12 per cent of young men drafted for active military service consume alcohol on a regular basis and 8 per cent have taken narcotics'. Furthermore, 'The educational level of soldiers is not high. About 7 per cent have only primary education, over 30 per cent do not have secondary education, and 40 per cent had not worked or studied before they joined the army.'[17] To make matters even more difficult, only '10.3 per cent of those eligible for military service will actually be called up in 2003'.[18] Why? For a number of reasons: deferments, medical problems (including families who bribe physicians to say they are ineligible), criminal backgrounds, drugs use, alcoholism, etc.

Finally, crime is widespread in the Russian military—and not just among enlisted personnel. Officers, including some very senior officers are also involved. In one case, for example, a general in charge of the supply troops in Siberia was charged with stealing more than one million roubles that was intended to purchase food for soldiers.[19] And the situation is not improving.

> In 2001 alone the Chief Military Prosecutor's Office recorded 1.1 per cent more crimes than in 2000. One in every two crimes committed by servicemen falls into the serious or particularly serious categories. Most often they are committed in a drunken state: The number of crimes of that kind increased 26 per cent in the year, while those involving weapons grew 25 per cent. The number of murders increased by an average of 7 per cent and instances of deliberating causing serious damage to health by 4.5 per cent.[20]

The question is why is the military faced with these problems? Is it just because of a lack of funds as some of the generals and admirals claim? Or is it an integral part of the kind of military structure that Moscow has at present? Certainly, there are serious structural problems. For example, NCOs (as the term is understood in the West) do not exist, and rather than get involved in dealing with junior enlisted personnel the way that the American or British NCOs do, officers avoid the problem. The result is brutality at the unit level.

These problems are reflected in the Russian military's operational effectiveness. Indeed, by most estimates, the performance of the Russian armed forces in Chechnya has been miserable, characterized by corruption, brutality, and an inability to bring this 'war' to a close. So what does this mean insofar as Putin is concerned? To suggest that he is not aware of the problems noted above would be naive. He is very aware of them, and he knows that something must be done to deal with them. He cannot afford to ignore and starve the military the way that Yeltsin did. It is important to him both as a force for protecting the country, as well as an influential political institution within the Russian polity.

Putin and Structural Changes

If there is a term to describe Putin's approach for dealing with the military's problems it is 'muddling through'. Putin has avoided the glossy frontal assault on the military's problems that some of his critics have suggested he follow.[21] Instead, as he has in other areas, he has overseen a number of small steps, culminating in his approval of the hugely important Military Reform Plan on 1 June 2003. To begin with, the Russian military now has only three services where it once had five. Second, the size of the Russian armed forces shrank drastically under Yeltsin. The military that Russia inherited from the USSR numbered 2.7 million. By 1 January 2001 this figure had been reduced to 1,365,000. In addition, the 600,000 troops that were stationed outside of Russia were brought back to the homeland, causing very serious dislocation problems—in particular in relation to housing.

Putin has adopted a cautious approach in dealing with the military. On 27 May and on 17 August, he set up commissions to look at problems within the armed forces. Their findings were placed on the agenda at Security Council sessions in August and November of 2000. The proposals were approved by the Security Council and dealt with the development of the military up to 2010. They included a broad range of issues, improving management of the military, increasing combat readiness, restructuring the defence industrial complex, upgrading the status of servicemen, increasing financial assistance to the military, and improving command and control.[22] The key point in comparing Putin's approach with the various reform plans that were put forward during Yeltsin's time is that the 2000 plan read more like a laundry list of items that needed to be fixed. There was no grand master plan. Instead, like the problem solver he is, Putin preferred to look at specific problems and see what could be done to deal with them one at a time.

Putin then closed down two of Moscow's three overseas bases—Lourdes in Cuba, and Cam Rahn Bay in Vietnam. When asked why the government chose to pull back by closing these bases, even though some senior military officers were opposed, Defence Minister Ivanov replied that they were too expensive:

> The use of the Lourdes base cost us 200 million dollars a year. It is true that Cuba owes us 20 billion dollars but when we raise the question they look surprised, pretending not to understand. Cuba will never repay this debt. But we kept pumping money into the Lourdes base and what did we get from it? The base was built in the 1960s as a tracking station. It was a different world and the base brought practical information results. Almost a half century has passed. A logical question arises: Do we need today what was good 40 years ago? And do we need it for $200 million a year? ... As for Cam Ranh Bay, in the past 15 years everything that could be stolen there was stolen and the rest broke down. The base was actually used for refueling our ships but they called at it only two or three times in the past ten years.[23]

The important point to note here is that there was no effort on Ivanov's part to explain Moscow's decision as part of a grand design. Instead, he suggested that Putin was given a cost-estimate analysis that indicated that Moscow was not getting its money worth—so the bases were closed.

However, Putin and his administration were looking at military reform in isolation and were savvy enough to understand that the military did not live in isolation. Military reform referred to 'a set of political, economic, legal, strictly military, military technical, social and other measures carried out at certain times and aimed at a quality transformation of the state's military organization'.[24] In other words, whatever was done in the military had to be seen against the backdrop of Russian society as a whole. But a grand plan? In reality Putin did accept an interrelated framework for military reform, but only after a number of bureaucratic battles and compromises had produced something that made sense as a strategy for bringing the Russian military into the twenty-first century.

On 11 May 2001, a commission that Putin had formed presented him with proposals on social problems faced by military personnel. Based on this report, Putin placed a priority on reforming the pay and allowances system as well as pension reform and improving housing and medical services. In October, Putin chaired a session of the Security Council that focused on the military-industrial complex. This resulted in an increased focus on the need to reform the Russia's military industry from the old Soviet model to a new, more competitive one. Finally, on 27 November 2001, Putin chaired another Security Council meeting that looked at the issue of mobilization readiness. It is also worth noting that in 2001, Putin succeeded in raising the military budget to 380.5 billion roubles. Another 34.6 billion roubles was allocated for pay and allowances as well as repayment of the Defence Ministry's outstanding debts.[25] The point here is that throughout the year, Putin pushed the bureaucracy to look at a wide variety of military related issues. He was clearly *not* ignoring the topic.

Given his tendency to work within the various bureaucracies, it was not surprising that Putin waited for the military to provide him with a proposal for changing its form of technical support—a plan that was delivered on 1 November 2000. This was followed by another proposal to reform the military-educational complex in July of 2001. Suggestions on how to handle military personnel were sent to him as well. He approved them on 16 November 2001. It was also at this time that he ordered the military to come up with a plan to fill certain positions with servicemen serving on contract. The idea was to experiment with one or two units in order to 'determine more precisely the nature and scope of measures and the outlays necessary for converting them to the contract method of manning with servicemen'.[26] The generals agreed that a volunteer, professional military was in many ways preferable to the existing system when it came to combat. On the other hand, there were many who wanted to keep the old system. In particular, conscripts were able to be utilized for free labour, and many believed that conscription continued to play an important role in socializing young men and instilling patriotic feelings.

The Defence Ministry was not the only one to come up with a plan for reforming the military. The Union of Right Wing Forces under the leadership of Boris Nemtsov also had its own version. According to this plan, the army would be transformed into a professional military almost immediately. Pay would be increased significantly to attract contract soldiers who wanted to serve. At the same time, conscripts would serve only six months, and instead of working in the fields and building dachas for generals they would be trained as specialists and then sent to the reserves.[27]

The idea of contract service in Russia was not invented by the Putin Administration. Contract service began on 1 December 1992 and targeted soldiers with certain specialties who were offered two-and three-year contracts. The program was gradually expanded until on 11 April 1996, the Chief of the Russian Defence Ministry's Planning and Mobilization Directorate stated that there were 270,000 contract soldiers in the military, many of them the wives of officers who joined because it was so difficult to live on an officer's salary.[28] Unfortunately, this experiment was a failure. In fact, during the 1993–5 time period, some 50,000 contract soldiers resigned, primarily for financial reasons.[29] For example, a contract serviceman earned 278,000 roubles compared to the average wage in Russia in September 1995 of around 550,000 roubles, including supplements. The subsistence minimum per person in Russia at that time was 300,000 roubles, and in some regions it was two or three times that.[30]

This situation did not improve. For example, in 1997 the Army's Chief of Recruitment observed that given the dangerous conditions and low pay faced by many contract soldiers, a person who volunteered for contract service, 'would either be one of the long term unemployed or someone who has already poisoned his mind with alcohol'.[31] And even those who joined quickly left. According to the same officer, in 1997, 'some 30,000 contract soldiers had left the army so far' that year, while 'only 15,000 had enlisted'.[32] In February 2001 it was reported that '49.9 per cent of contract servicemen have an income lower than the official subsistence minimum'.[33] And it was clear that the army was still not attracting the kind of quality individuals it sought as professionals. Those who joined were often 'the most unfortunate layers of the population, that part least prepared for market conditions, and on the other hand, young people not easily adaptable to barracks life'.[34]

During his 1996 Presidential Campaign Boris Yeltsin promised to end conscription by the year 2000. However, as was often the case with Yeltsin, he failed to factor in the economic costs involved or follow through on his promises. Indeed, while a professional military may be popular among the populace, and while such individuals do a better job in combat, it is considerably more expensive than a conscript force structure. As one Russian general put it, 'A Conscript costs us 17,900 roubles a year, while a professional soldier costs 32,000 roubles. A professional army would require the corresponding infrastructure, which also would cost a lot. It is not [a realistic option] now.'[35] Or as another senior Russian General put it in 2001, 'because of the lack of funds it will not be possible for Russia to build a fully professional army for at least the next five years'.[36] The introduction of contract soldiers was controversial and opposed by most of the country's senior military officers. They complained (rightly) about the costs involved, and they worried about its impact on the combat readiness of the Russian Army. However, something had to be done, and it had to be more than a case of throwing money—which the country did not possess—at the military's problems.

Recognizing the bureaucratic difficulties he faced in dealing with the military, but convinced that he had to keep pushing the generals and admirals if he hoped to get anything done, Putin decided to carry out what was labeled an 'experiment'. The idea was to fully professionalize the 76th airborne division based at Pskov; to see how realistic the idea of professionalizing the military would be. The experiment began on September 2002 and was to last a year. Contracts would make it very clear that the contract serviceman would be expected to serve in difficult or dangerous combat areas such as Chechnya. In addition, the individual had a right to housing—something new for enlisted personnel in the Russian Army.[37] According to some observers, the military tried to 'sabotage' the experiment from the beginning.[38] For example, on 28 September 2002, Chief of the General Staff, General Anatoliy Kvashnin visited the Pskov division and announced that 'no one intended to give apartments to the soldiers'.[39] As a result, 48 soldiers, who had previously signed contracts, submitted requests for discharge.

The major problem faced by those who wanted the experiment to succeed, was how to make military service attractive. As far as the 76th division was concerned, the experiment was failing. It was supposed to attract 2,600 contract troops, but by the first part of October 2002 it had only recruited 749 candidates, and of them, 'four hundred passed the medical exam, and two hundred and fifty signed a contract'.[40] Indeed, by November, the experiment had stalled because of a lack of volunteers. The division was only able to attract about 600 young men—in a unit that required 1,630. Clearly something had to be done. In response to the situation in Pskov, the Union of Right Forces suggested that the pay for soldiers which was being held at 3,000 roubles per month should be raised to 4–4,500 roubles. Furthermore, they recommended that officer pay should be increased by 40–50%.[41] They further proposed that soldiers who signed a contract should be given free tuition if they decided to continue their education when their time of service was over, and that conscription abolished immediately.

Putin was well aware of problems such as those at Pskov. For example, the Russian government held a meeting on 26 November 2002. At this meeting, Putin stated that he had a 'generally positive' outlook on the changes underway in the military—a clear sign that he expected further movement toward a professional military. By this point, the generals were proposing that the transformation go through three stages. The first, the Pskov division, was to end in 2004. The second, according to defence minster Ivanov, was to cover a period of seven years. The goal was to have 50–60% of soldiers and sergeants serving on contract. The third stage was left undefined—'It is planned that all armed forces will be manned with "contractees" during the third stage.'[42] There were problems with this staged approach, however. In particular, the second stage would end after Putin had left office which many critics believed would give the generals the opportunity to simply wait the reform plan out in the hope that it would falter under a new administration.

Apparently feeling strong enough to take on the generals, Putin and Ivanov, began a campaign to bring together the outlines of a serious military reform program. First, Putin stated on 29 November 2002, that in the future Moscow would place primary reliance on professional soldiers, not conscripts, to fight the country's wars.[43] Then, speaking to journalists, Ivanov stated that priority would be given to recruiting individuals to serve in the airborne troops, the 42nd Motor Rifle Division as well as the Naval Infantry. He also outlined a proposed figure of 166,000 contract soldiers in the military as a whole. This 'applies to the 92 subunits in a permanent state of combat readiness which will form the basis of a professional army'.[44] For the first time, Ivanov was drawing the outlines of what would become a mixed force, but with some units dedicated to front line combat while others were committed to the reserves. By the end of 2002, commentators were suggesting—for the first time—that the government might have turned the corner in introducing changes in the military. However, it was obvious that even 4,000 roubles a month would not be enough to attract soldiers to serve on contract. Putin promised to address the pay issue and to continue pay raises until military service became attractive.

April 2003 was a key month for reform in the Russian military. On 24 April, the government endorsed a long-term reform plan that clearly demanded compromise from the generals. The structure of this reform plan can be divided into five parts.

1. *Conscription—from two years to one?* The most controversial proposal was how long conscripts would be obliged to serve. Conscription remains deeply unpopular amongst draft-age men in Russia, but the military needs manpower. Whether or not conscript service be limited was therefore a critical issue for the Kremlin. The bottom line was rather simple. The Duma provided so many ways to get conscription deferments that only around 11% of eligible Russian men were eligible to be drafted. Moreover, many of these could not be drafted because of poor health, or drug use or criminal background. In addition, draft evasion has become endemic. The initial defence ministry reform plan represented a move toward the kind of military Putin desired. For example, Ivanov suggested that the Defence Ministry was 'prepared to consider the question of reducing the length of compulsory military service from two years to one'.[45] Assuming such a program were adopted, the first six months would be spent at a training center. During the second six months the conscripts would be attached to a regular unit. These would not be the 'permanent-readiness' units which will be made up of professionals, but second line forces. By the end of April, the Deputy Chief of the General Staff himself stated that from 2008 service time would be reduced to 12 months. At the end of six months conscripts would have an opportunity to sign a contract. It was clear that Putin came down hard on the military on this issue, and forced them agree to an explicit date (while he was still in office) on which the old conscription system would end.

2. *A Bifurcated Army?* Assuming the April 2003 plan is actually implemented, Russia will have two different armies. The first 'army' will be made up of contract soldiers. Within five years, Ivanov stated that 92 units made up of ground forces, airborne troops and marines will be all-volunteer. All together this will mean 10 divisions, seven brigades and 13 regiments; a total of 166,000 soldiers. This process will start in 2004.[46] Units serving in the North Caucasus will be transferred to contract soldiers during 2004–5. Then during the period from 2004 to 2007 the entire military will be modified so that those units on 'permanent standby' and those which 'substantially affect the combat readiness of the country' will become fully staffed with soldiers serving on contract [47] The rest of the military would be made up of junior officers who would train those conscripts who decided to leave after the first year. In essence, Russia would have a front line military force capable of being deployed at a moment's notice and a reserve force useful in the event of a national emergency.

3. *A Foreign Legion?* Given how desperate the Kremlin was for manpower, it decided to reach out to citizens (mostly Russians) living in the Commonwealth of Independent States (CIS). Individuals who volunteered to serve in the Russian military will be given accelerated citizenship (they will only have to wait three instead of five years) if they serve as contract soldiers. Priority will go to those individuals who have 'military professions or previous military training'.[48] All CIS volunteers will be integrated with regular Russian troops, and the language of command will be Russian.

4. Women in Combat? In the past, women in the Russian armed forces have been relegated to support roles such as medical or administrative positions, although some have been permitted to serve in units such as the Strategic Rocket Forces.[49] The Russian Army today is opening up positions for women which will permit them to serve 'in combat roles'. Given their traditional resistance to this approach, this move is a clear indication of just how desperate the military's manpower problem has become.

5. NCOs? The Russian government understands on one level that it is in need of the kind of NCOs that exist in the West and as part of the military reform plan, they intend to institute special training for them. The hope is that with a well-trained NCO corps the army will be able at long last to do something about its hazing problem. However, while there are clear advantages to introducing an NCO corps—not least that it will require officers to delegate authority for the first time—it will also require higher salaries. For example, at present conscripts make about $3 a month. Low ranking officers make an average of $75 per month, while an estimated 20,000 non-commissioned officer positions remain unfilled.[50] In addition to raising salaries and delegating authority, the Russian military would also have to make major changes in its over-bloated officer corps.

This military reform process would not be cheap. The cost of transforming the military to this mixed professional/conscript forces is estimated to be around 135 billion roubles ($4.34 billion), with the majority of this money going to paying wages and creating the kind of infrastructure (such as housing, schools, medical facilities) that is critical to the maintenance of a professional military.[51] In order to ensure that sufficient funds to pay for this experiment are on hand, Putin set money aside in the 2004 budget. However, to get an idea of how difficult this problem will be, in 2003 alone, 14.7% of all government spending (333 billion roubles) was allocated to national defence—and this figure did not include the cost for the internal security forces.[52] If the two are taken together, fully one-third of Russia's state spending is allocated to defence and security. As a consequence, finding an additional 135 billion roubles will not be easy. Indeed, on 24 April 2003, Prime Minister Mikhail Kasyanov approved only 30bn roubles of Ivanov's 138bn rouble supplemental budget request.[53]

In spite of these actions, Boris Nemtsov and his party continued to challenge the Putin on the military reform issue. Nemtsov told Putin that if the military did not accept his party's proposal (in lieu of the longer, phased approach advocated by the military), his party 'will work to obtain the resignation of Defence Minister Sergey Ivanov'. Putin's response was to suggest 'to Nemtsov that he calm down and keep trying to find a common language with the defence minister'.[54] During this meeting—which was attended by a wide variety of government officials—senior military officers argued that Nemtsov's plan would radically reduce the size of the military in two years—and leave it without a new recruitment mechanism.

Illustrating the importance he attached to it, Putin made military reform one of the three issues he singled out for special attention in his state-of-the-union address on 16 May 2003. To begin with, he noted that Moscow 'will continue to form a permanent, professional rapid reaction force'. As far as conscription was concerned, Putin was categorical, observing that it would be reduced from two years to one in 2008. He also reiterated the government's plan to provide preferential entry into Russian institutions of higher education for those who serve three years, and he repeated Ivanov's promise to grant accelerated citizenship to individuals from the CIS countries who served in the Russian military.[55]

Attacks on the General Staff

One of the techniques that Putin successfully used to push the military in the direction he wanted—if not to change their attitudes—has been to seize on events to move matters in the direction he favored. The first such event was the Attack on the Twin Towers in New York City.

Of all the events that took place in 2001, few shook the Russian military more than the attack in New York. While Putin may be a turtle in his tendency to push matters through the bureaucracy, he has shown that he is also able to seize on such situations and use them to his advantage. This is exactly what he did in this instance. Putin believes very strongly that good relations with the United States are key to Russia's economic and political recovery and the attack on Twin Towers was a perfect opportunity to improve US–Russian relations. He was the first foreign leader to call President George Bush, and he made it clear that he was prepared to help in any way. The Russian military, however, was not particularly forthcoming and their suspicion of the West and NATO remains strong. As a result, Putin called the senior generals together on 24 September, just before he traveled to Bush's ranch in Crawford, Texas. At this meeting, Putin informed Moscow's generals and admirals that it was time to recognize that there had been a major change in the country's security policy. And the generals got the message. Shortly afterwards, the Kremlin dropped its opposition to the presence of US forces in Central Asia, and became very helpful in providing the US with intelligence information relative to the conflict in Afghanistan. Furthermore, the Defence Ministry toned down its opposition to initiatives such as greater cooperation between Russia and NATO.

The situation was similar when a large helicopter—that was clearly overloaded—crashed and killed all on board in Chechnya which led to a number of high profile dismissals. Putin also seized on terrorist attacks within Russia and elsewhere and demanded that the Russian armed forces adapt themselves to be able to fight terrorism on a global scale. As Putin put it, terrorism 'is increasingly becoming a factor of global policy and, therefore, the Armed Forces should oriented primarily to oppose this phenomenon'.[56] In practice this meant the army had to 'develop operational plans and train personnel to secure important objects and to storm them if they are captured by terrorists'.[57] In fact, a Russian general writing in October 2002 underlined just how out of date the Russian military was:

> The most effective way to fight terrorists in cities is not the broad-scale military operations that are being conducted in Chechnya today; it is focused work by agents to identify and forestall terrorist acts. The main thing is to obtain information about preparations for terrorist operations in time. Of course, if there had been such information, measures would have been taken. But you cannot assign special forces detachments to every theater. Therefore the role of all the special structures who are getting information about the terrorists and their plans unquestionably increases.[58]

General Kvashnin, the Chief of the General Staff, followed up a month later, ordering all 'military units and garrisons to submit contingency plans for the prevention of terrorist attacks'.[59] From a bureaucratic standpoint, this meant that Putin had the generals and admirals just where he wanted them. He had come up with a new task, and they were at a loss on how to respond because their military was not structured to deal with this new threat. They would have to change structure and when that happened it would become more difficult to oppose the President on other issues.

The Allied invasion of Iraq strengthened Putin's position against the generals still further. Indeed, the speed and effectiveness of the American assault on Iraq stunned the country's military leaders. As Pavel Felgenhauer has noted:

> Russian generals were expecting another prolonged so-called non-contact war, like the one against Yugoslavia in 1999, in Afghanistan in 2001 or the first gulf war in 1991, when a four-day ground offensive was preceded by a 39-day air bombardment. It was believed that the Americans were afraid of close hand-to-hand encounters, they would not tolerate the inevitable casualties.[60]

The situation looked more bleak from the military's standpoint when it became known that three Soviet period generals had visited Baghdad prior to the American attack and purportedly given the Iraqis advice on how to resist the American attack. The ineffectiveness of this advice suggested that the Russian military had underestimated how far behind the Americans technologically it was. As far as Putin's relationship with the military was concerned, this event was clearly an argument for major reform. 'The Russian army should immediately draw lessons from the Gulf War ... A key conclusion that we should draw is the obsolete structure of the Russian armed forces must be changed.'[61] Or as a member of the Duma put it, 'The Iraqi war has proven once again that a volunteer contract force equipped with state-of-the-art weapons and using modern tactics can fulfill any task.'[62]

These developments made it very clear that change was a necessity, and permitted Putin to push the issue through at the 24 April meeting. On 10 July 2003 the cabinet formally approved the Defence Ministry's reform proposal. If the plan is implemented on time, around 49% of the Russian military will be serving on contract by 2007 while the time served by conscripts will be lowered to one year. In a press conference at the time, however, Defence Minister Ivanov warned that a fully professional military 'was still many years away'.[63]

Faced with Kvashnin's many schemes and efforts to derail Ivanov's efforts to reform the military, Putin and Ivanov came to the conclusion that they had to do something. Kvashnin was not only devious, he was reportedly a very difficult individual to get along with, a view shared by a large number of senior officers. However, the problem went deeper than Kvashnin's personality. It was structural as well. According to the *Defence Law*, responsibility for defence affairs was split between Ivanov and Kvashnin. Under this law, the Chief of the General Staff and the Defence Minister were equal when it came to decision-making power on military related issues. It 'was as if chief and subordinate were placed on the same level. This "elevation" strongly suited Army General Kvashnin, but it caused turmoil in the control of forces. Matters reached the point where a document signed by the minister but lacking the General Staff's Endorsement was sometimes not acted on.'[64] This was especially true of operational matters where Kvashnin took a special interest. Something had to be done.

Having given up on efforts to work with Kvashnin, Ivanov fired the opening salvo in January 2004 at the annual conference of the Military Sciences Academy, Ivanov openly criticized the General Staff arguing that it 'spends too much time on superfluous administration and day-to-day management of the troops, to the detriment of its main purpose: situational analysis and development of troop deployment plans'. As a consequence, he maintained, 'good ideas are by no means always converted into the best results'.[65] It was clear to everyone that Ivanov would not have made such a charge if he did not have Putin's support and if he were not prepared to follow it up with action.

Action was not long in coming. A couple of months later special legislation was introduced. If adopted this bill would fundamentally change the relationship between the Ministry of Defence and the General Staff. To be specific—the old law had read, 'oversight of the Armed Forces of the Russian Federation is carried out by the defence minister via the Defence Ministry and the General Staff of the Armed Forces, which is the main body of operational supervision of the Armed Forces'.[66] However, the new language omitted any mention of the General Staff. Furthermore, recognizing the reduced size of the Russian Army (compared with 1990, for example), the legislation took away two of the deputies previously authorized for the Ministry of Defence—now there would be only four. In addition, the General Staff would now assume the position

Ivanov believed appropriate: it would become what the now legendary Chief of the General Staff under Stalin, Boris Shaposhnikov had called 'the Brain of the Army'. It stayed completely out of operational matters. In short, when this bill was adopted in mid-2004, Ivanov was placed in charge of all aspects of military affairs, including operational matters. To wit,

> This decision will put an end to the 'four-year war' for the right to control the army waged by the General Staff and the Defence Ministry and will be one of the elements of ongoing reform of federal organs of power. From now on only one person, Defence Minister Sergey Ivanov, will be entirely responsible for decisions on issues of the country's Armed Forces organization.[67]

Meanwhile, Putin and Ivanov seized on a surprise attack by Chechen rebels in Ingushetia which reportedly killed 100 and wounded another 100 to carry out a purge of top officials in the military and security agencies—including General Kvashnin.[68] He was replaced by General Yuri Baluyevsky, a highly regarded officer who spent the majority of his career in staff positions, and somebody who—in contrast to Kvashnin—is more likely to follow Ivanov's instructions to limit the General Staff's work to areas such as military reform and strategy.[69] The tasks facing Ivanov are significant. By the 2008 election 'he needs to reduce the period of conscript service to one year, to improve the permanently failing system of service under and to implement a mortgage accumulation scheme of housing provision to servicemen'.[70] One might also add, that it will be up to him to come up with a meaningful reform plan—a task that has bedeviled Russian decision makers every since Mikhail Gorbachev came to power. The advantage, however, is that he will no longer have to compete with the Chief of the General Staff, and in addition he now has 'a central staff (unprecedented in either Tsarist or Soviet military organizations) numbering around 9000 personnel'.[71]

Conclusion

Will the changes outlined in Putin's military reform program and the changes made in the relationship between the General Staff and the Ministry of Defence be enough to create a new kind of armed forces? It is difficult to say for certain, but there is no doubt that a major change has taken place. If any general or admiral has any question about who is in charge of military affairs, one suspects that Ivanov is now in a position to answer it—he is.

Having noted the important changes that have taken place, it is also important to keep in mind some of the other difficult problems that face Ivanov. For example, the experiment at Pskov was far from successful and the division still lacks around 2000 volunteers. The recruitment plan is only 22.5% completed. Moreover, there are continuing difficulties inside the unit itself. 'For

62 per cent of service members, expectations from contract service had not been met, and for 91 per cent of these it was owing to the low level of pay and allowances ... In addition ... 72 per cent of service members are dissatisfied with the social and living conditions of service, partly because of the lack of housing.' This led Colonel General Georgiy Shpak to ask that the recruitment period which was supposed to end 1 May 2003 be extended to 1 October 2003.[72]

There are also problems associated with the attitudes of many military personnel toward the country's leadership. A recent survey by the National Public Research Center (VtsIOM) of professional soldiers in Moscow (conducted in May 2003) indicated that the 'military does not trust its leaders; none of the present heads of the Defence Ministry and General Staff won the approval of more than 5 per cent of the respondents'.[73] Moreover, the institution of an NCO corps in the military is fraught with difficulty. For many Russian officers, the idea of delegating authority to the degree that Americans and British do is an uncomfortable prospect. Thus, when Russian naval officers visited an American ship in Norfolk some years ago, the Commander of the Northern Fleet asked how many men serviced a particular missile mount on board a US ship. When he was told that a 20-year-old petty officer second class was in charge of both maintenance and firing the weapon, the Russian could not believe what he was hearing. In the Soviet Navy, two officers would be used, one to maintain the weapon and one to fire the weapon. Yet this kind of delegation of authority is critical if the Russians hope to attract volunteers to man their NCO corps.

One of the key questions facing Putin is the old one—money. His critics are right, he must raise salaries if he hopes to attract volunteers. Furthermore, he must provide the 'creature comforts' that are so much a part of a professional, volunteer military. So far, the military does not have the 138 billion roubles it needs to implement the program.

So where does that leave military reform in Russia? Putin has pushed the military further toward reform during the last year than in the previous 12 years since the collapse of the USSR. It is always possible that the military will outlast him as many have suggested. However, if the April 2003 plan (which was approved in June and endorsed by the Duma in July) is implemented, the Russian military will have a different structure—a professional military backed up by a reserve force. In this regard, it is important to again emphasize that the Russian military is being pushed hard to change by events such as Iraq, the new fight against terrorism, and recognition on its own part that it can no longer compete in the world of modern combat. A military made up primarily of conscripts cannot fight the kind of war the US fought in Iraq. If it wants to be competitive with Western armed forces, the Russian military must change. Present indications suggest that Putin will continue to push the issue until the military is ready to make the necessary changes.

Notes

[1] Lilia Shevtsova, *Putin's Russia* (Washington, DC, Carnegie Endowment for International Peace, 2003), especially chapter 3.

[2] Lilia Shevtsova, 'Putin's Dilemma: To Stabilize or Transform?', *Moscow Times*, 13 January 2003 in *Johnson's List*, 14 January 2003.

[3] Dale R. Herspring (ed.), *Putin's Russia, Past Imperfect, Future Uncertain* (Boulder, Roman and Littlefield, 2002), pp. 259–262.

[4] See for example, Stephen Blank, 'This Time We Really Mean it: Russian Military Reform', *Russia and Eurasian Review*, 2: 1 (7 January 2003); Roger N. McDermott, 'Putin's Military Priorities: Modernization of the Armed Forces', *Insight*, 3: 1 (2003); Dale R. Herspring, 'De-Professionalising the Russian Armed Forces', in Anthony Forster, Timothy Edmunds & Andrew Cottey (eds), *The Challenge of Military Reform in Postcommunist Europe: Building Professional Armed Forces* (Houndmills: Palgrave-Macmillan, 2002), pp. 197–210; Dale R. Herspring, 'Putin and the Armed Forces', in Herspring, *Putin's Russia*, pp. 155–175; Peter Rutland, 'Military Reform Marks Time', in Peter Rutland, 'Russia in 2002: Waiting and Wondering', *Tansitions Online*, available at http://www.fol.ts.

[5] There are a number of articles and chapters that discuss these problems. See for example, Zoltan Barany, 'Politics and the Russian Armed Forces', in Zoltan Barany & Robert G. Moser, *Russian Politics: Challenges of Democratization* (Cambridge, Cambridge University Press, 2001), pp. 175–214, Herspring, 'De-Professionalizing the Russian Armed Forces'.

[6] Yelena Shishounova, 'Defence chiefs briefs Putin on state of military', 9 January 2003, gazeta.ru in *Johnson's List*, 9 January 2002, p. 12.

[7] 'Russian Chief Army Prosecutor on Bullying, Draft Problems, Bribes, Specific Cases', *Rossiyskaya Gazeta*, 6 June 2003, available at http://www.dialogselect.com/intranet/cgi/present (accessed 13 June 2003).

[8] 'Chief of General Staff urges more money, training for Russian armed forces', *Rossiyskaya Gazeta*, 14 May 2003 in *Johnson's List*, 15 May 2003, p. 22.

[9] 'Chief of General Staff urges more money', p. 24.

[10] 'Russian Air Force command worried about aging of aircraft', *ITAR-TASS*, 26 March 2003, in Wnc@apollo.fedworld.gov, *WNC Military Affairs*, 27 March 2003.

[11] 'Russian Military Continues to Decay While Few Projects are Funded', *Trud*, 18 January 2003, in Wnc@apollo.fedworld.gov, *WNC Military Affairs*, 11 February 2003.

[12] 'Admiral Komoyedov on Future of Russian Navy, Views on CINC Navy Kuroyedov, *Tribuna*, 30 October 2002, in Wnc@apollo.fedworld.gov, *WNC Military Affairs*, 30 October 2002, p. 4.

[13] 'Defence Ministry Says Military Procurements To be Cut in 2004', *The Monitor*, 11 August 2003.

[14] 'Russian Expert Mulls Political, Social Reasons for 'Never-Ending' Army Desertion', *Rossiyskaya Gazeta*, 19 June 2002, in WNC@apollo.fedworld.gov, *WNC Military Affairs*, 20 June 2002, p. 1.

[15] Anna Badkhen, 'Hard Times for Russia's Reeling Military', *San Francisco Chronicle*, 2 December 2003.

[16] 'Vladimir Georgiyev, 'The General Staff Releases Crime Statistics', *Nevavisimaya Gazeta*, 12 July 2002, in *Johnson's List*, 12 July 2002, p. 14.

[17] 'Russian Mps say army desertion mirrors criminal trends in society', *ITAR-TASS*, 18 June 2002, in WNC@apollo.fedworld.gov, *WNC Military Affairs*, 19 June 2002.

[18] 'Only 10.3 per cent of Those Eligible for Military Service Will Actually be Called up', *Rosbalt*, 10 April 2003 in *Johnson's List*, 10 April 2003, p. 25.

[19] 'Security Council Amending Military Reform Plans Again', *Nezavisimaya Gazeta*, 4 June 2002, in *Johnson's List*, 8 June 2002, p. 25.

[20] . 'Russia: Former Military Prosecutor Cited on Army's Ills', *Izvestiya*, 15 June 2002, in WNC@apollo.fedword.gov, *WNC Military Affairs*, 18 June 2002.

[21] For example, Pavel Felgenhauer suggested to this writer in 2002 that Putin could do anything with the military he wanted, 'he is the president, and they will obey'.

22. See, 'Russia: Survey of Military Reform in the Russian Federation', *Yadernyy Kontrol*, 19 April 2002, in WNC@apollo.fedworld.gov, *WNC Military Affairs*, 6 September 2002, p. 5.
23. . 'Defence Mnister Sergei Ivanov About Iraq, UN, NATO and the Russian Army', *Komsomolskaya Pravda*, April, 2003, nos. 58–59, in *Johnson's List*, 3 April 2003, pp. 27–28.
24. 'Russia: Survey of Military Reform', p. 1.
25. 'Russia: Survey of Military Reform', p. 8.
26. 'Russia: Survey of Military Reform', p. 10.
27. 'Major Clash Shaping Up Between Rightest Politicians and Military Chiefs on Military Reform', *Obshchaya Gazeta*, 13 December 2001, in WNC@apollo.fedworld.gov, *WNC Military Affairs*, 17 December 2001.
28. 'Staffing Levels in the Russian Army', *The Monitor*, 11 April 1996.
29. 'Conversations Without Middlemen', *Moscow Television*, 18 September 1995, in *FBIS*, CE, 18 September 1995, p. 21.
30. *Krasnaya zvezda*, 12 September 1995.
31. 'Russian Army Struggles with contract Service', *The Monitor*, 15 October 1997.
32. 'Russian Army Struggles with contract Service.'
33. 'Soldiers Face Major Housing Shortages', *RFE/RL Daily Report*, 24 February 2001.
34. 'More on Russian Defence Concept: Military Reform', *The Monitor*, 14 August 1998.
35. 'Russia's Army Still Mired in Conscript Crisis', *The Russia Journal*, 24–30 April 2000 in *Johnson's List*, 27 April 2000.
36. *The Monitor*, 27 June 2001.
37. 'Russia: Problems, Including Low Pay, in Converting 76th Airborne Division to Unit of Contract Soldiers', *Rossiyskaya Gazeta*, 16 August 2002, in *Johnson's List*, 21 August 2002, p. 2.
38. Pavel Felgenhauer, 'Leaking, Lobbying, Looting', *Moscow Times*, 18 July 2002 in *Johnson's Report*, 18 July 2002, p. 5.
39. 'Russia: Scandal Brews Over Military Reform, Pskov Contract Service Experiment'. *Moskovskiye Novosti*, 8 October 2002, in wnc@apollo.fedworld.gov, *WNC Military Affairs*, 10 October 2002.
40. 'Litovkin: "political games" doom contract service experiment in 76th Division', *Vremya*, 8 October 2002, in wnc@apollo.fedworld.gov, *WNC Military Affairs*, 24 October 2002.
41. 'Litovkin: "political games" doom contract service experiment.'
42. 'Russian defence minister to address government on military reform', *ITAR-TASS*, 20 November 2002, in wnc@apollo.fedworld.gov, *WNC Military Affairs*, 21 November 2002.
43. 'Russia's Putin Says Professional Soldiers, Not Recruits Should Fight in "Hot Spots"', *Moscow Interfax*, 29 November 2002, in wnc@apollo.fedworld.gov, *WNC Military Affairs*, 2 December 2002.
44. 'Russian defence minister outlines professional army timetable', *ITAR-TASS*, 20 December 2002, in wnc@apollo.fedworld.gov, *WNC Military Affairs*, 23 December 2002.
45. 'Russia Considers Reducing Conscription From Two Years to One', *ITAR-TASS*, 24 April 2003, in wnc@apollo.fedworld,gov, *WNC Military Affairs*, 25 April 2003.
46. Simon Saradzhyan, 'Core of Army to Go Contract by 2008', *Moscow Times*, 22 November 2002, in *Johnson's List*, 22 November 2002.
47. 'Government sides with generals over military reforms', *gazeta.ru*, 24 April 2003 in *Johnson's List*, 24 April 2003, pp. 11–12.
48. 'Defence Ministry Sets Requirements for Contract Soldiers from CIS', *ITAR-TASS*, 22 May 2003, in wnc@apollo.fedworld.gov, *WNC Military Affairs*, 28 May 2003.
49. See Dale R. Herspring, 'Women in the Russian Military: A Reluctant Marriage', *Minerva*, 15: 2 (Spring, 1997), pp. 42–59.
50. 'Russia recruits army volunteers', *Atlanta Journal-Constitution*, 5 January 2003, in *Johnson's List*, 5 January 2003, p. 15.
51. Simon Saradzhyan, 'Army's Plan for Reform Wins Out', *Moscow Times*, 25 April 2003 in *Johnson's List*, 25 April 2003. Other estimates went higher, for example to 138 roubles. 'Russian defence minister on military reform', *ITAR-TASS*, 25 April 2003, in wnc@apollo.fedworld.gov, *WNC Military Affairs*, 27 April 2003.

V. Putin and Military Reform in Russia   155

52 'Russia Gives Priority to National Defence in 2003 Budget', *Interfax*, 13 June 2002, in *Johnson's List*, 13 June 2002. Apparently, the budget was raised to 344 billion roubles.

53 'Government Slashes Military Chiefs Reform-Funding Request', *RFE/RL Report*, 25 April 2003.

54 'United Russia Puts Active Support Behind Military Reform', *Tribuna*, 16 May 2003, in wnc@apollo.fedworld.gov, *WNC Military Affairs*, 28 May 2003.

55 'Russia pledges to follow through on delayed military reform', *AFP*, 16 May 2003, in *Johnson's List*, 17 May 2003.

56 'Russia's Putin Says Armed Force Should Focus on Opposing Terrorism', *ITAR-TASS*, 26 November 2002, in wnc@apollo.fedworld.gov, *WNC Military Affairs*, 27 November 2002.

57 'Defence Minister to Prepare Army for War Against Terrorism', *RFE/RL Report*, 5 November 2002.

58 'Russian General, Senator Reviews Needs in War Against Terrorism', *Rossiyskaya Gazeta*, 31 October 2002, in wnc@apollo.fedworld.gov, 7 November 2002, p. 2.

59 'General Staff Orders Troops to be Ready for Terrorist Attacks', *RFE/RL Report*, 11 December 2002.

60 'The Elite's Feeling the Heat', *Moscow Times*, 10 April 2003 in *Johnson's List*, 10 April 2003.

61 BBC Monitoring, 'Russian army reform should draw on Gulf war lessons—expert', 10 April 2003, in *Johnson's Report*, 10 April 2003, p. 23.

62 Maria Golovnina, 'Russia Speeds up Army Reform, Analysts Skeptical', 15 April 2003, *Reuters*, in *Johnson's List*, 15 April 2003.

63 'Russia Creeps Slowly Towards Fully Professional Army', *AFP*, 10 July 2003, in *Johnson's List*, 14 July 2003.

64 'Russian Defence Minister: General Staff Shedding "Inappropriate Functions"', *Rossiyskaya Gazeta*, 15 June 2004, in *WNC Military Affairs*, 17 June 2004.

65 Oleg Odnokolenko, 'Decisive Battle', *Itogi*, 3 August 2004, in *Johnson's List*, 5 August 2004.

66 Valery Bashkirov, 'A Staff Operation Against Kvashnin', *Russkii Kurier,* 29 April 2004, in *Johnson's List*, 29 April 2004.

67 'Lowering of General Staff Status Expected to Put an End to Ivanov-Kvashnin "War"', *Izvestiya*, 10 June 2004, in *WNC Military Affairs*, 16 June 2004.

68 The actual numbers vary depending on the source. The main point, however, is that a very large number of soldiers and civilians were killed and wounded, and as the 'operational' commander of the Russian Army, Kvashnin as well as several other Army and Internal Security officers were responsible.

69 This writer had an opportunity to observe General Baluyevsky in action as a member of the party accompanying Admiral Dennis Blair (Commander in Chief, US Forces Pacific (CINCPAC)) on his visit to Russia in 2002. He impressed everyone as a highly organized, intellectually sophisticated, well-prepared officer able to discuss a wide variety of issues without the use of briefing books.

70 'Lowering of General Staff Status'.

71 'Lowering of General Staff Status'.

72 'Lessons learnt from Russian armed forces contract experiment—defence minister, *Krasnaya zvezda*, 29 May 2003, in *Johnson's List*, 2 June 2003, p. 34.

73 'The Military Doesn't Like its Leaders', *Nezavisimaya Gazeta*, 4 June 2003, in *Johnson's List*, 4 June 2003, p. 14.

Ukraine: Reform in the Context of Flawed Democracy and Geopolitical Anxiety[1]

JAMES SHERR
Conflict Studies Research Centre, Defence Academy of the United Kingdom, Camberley, UK

Ukraine's Armed Forces are undergoing in-depth reform in a context of pervasive uncertainty. In significant measure, that uncertainty is socio-political. The Armed Forces are a core structure of a state in which power is grossly weighted towards a president who is widely distrusted by the country's citizens. Whatever reform has been achieved has depended upon the support, or at least the acquiescence of Leonid Kuchma and those who have gained his favour. The merits of what has been achieved, like so much else in Ukraine, have not been communicated to the population—which, not surprisingly, perceives that there has been little or no defence reform at all.[2] Civil society, whilst far from absent in Ukraine, is fragmented and resentful, and it has far more immediate and consuming priorities than the effectiveness and well being of the Armed Forces.[3] In view of the fact that the President is required by the current (1996)

Constitution to leave office and hold elections in autumn 2004, this is not a trivial matter. For the centre-right political opposition—which commands at least a plurality of support in society—the consuming priority is Ukraine's democratisation. In neither of the centre-right blocs are there many who have ties to the military establishment, let alone substantial sympathy for it. This, too, is not a trivial matter. Should the 'regime' survive under another name, the Armed Forces are likely to remain hostage to the intentions and whims of Kuchma's successor. Should a form of regime change take place, then the uncertainties of the Armed Forces are likely to be greater than they are now.

Yet the uncertainty is also geo-political. Defence reform has become the centrepiece—and the day to day business—of Ukraine's relationship with NATO: the one relationship between Ukraine and the West which, in the eyes of the country's authorities, has brought practical benefit to Ukraine. These authorities have used this relationship for both political and geopolitical advantage: politically, to muffle Western reactions to growing authoritarianism; geopolitically, to offset—and persuade the West that it is *determined* to offset—very powerful pressures from Russia and a powerful alternative dynamic of integration in the former USSR. But nothing is guaranteed in the NATO–Ukraine relationship. In the post-9/11 world, NATO has more pressing and more dire preoccupations than Ukraine. Whilst the mechanisms of working level cooperation between NATO and Ukraine are eminently sustainable in this new world, the political will required to move this relationship forward from 'distinctive partnership' to accession may not be forthcoming. The only fresh elements in this relationship in recent months are disturbingly asymmetrical: courageous steps by Ukraine to support coalition activities in Iraq—activities that fall outside NATO but which are critical to its future—and renewed efforts by NATO to underscore the importance it attaches to democratisation in Ukraine. Thus, at a time when much of Ukraine's opposition perceives that NATO has underwritten an undemocratic regime, elements inside that regime perceive that it is exerting internal political pressure without offering any geopolitical compensation or reward. For at least ten years, Ukraine's 'full integration into Euro-Atlantic security structures' and defence reform have been indivisible pursuits. Should the two pursuits become decoupled, defence reform may lose its most powerful constituent, and reformers may find themselves marginalised by authorities who have marginalised so many others.

Civil–military Relations

Any assessment of relations between society, political institutions and the Armed Forces is bound to be misconceived if it is not informed by a historical perspective. The Soviet Armed Forces was not only a military machine, but a social institution, which has left behind deeply entrenched attitudes about authority, society, national security and the role of the military in defending it.

Its legacy is by no means entirely bad, but it is profoundly influential, and it both shapes and hinders progress.

In the USSR political control over the Armed Forces was at one and the same time pervasive, but narrowly focused. Through the Defence Council, the Chief Political Directorate of the Communist Party Central Committee and the 'special departments' of the KGB, the Soviet Politburo had mechanisms at its disposal which not only ensured the reliability of the Armed Forces, but their total obedience. Paradoxically, the very effectiveness of these mechanisms persuaded the Party leadership to entrust the Armed Forces with a dominant influence in military-technical decisions and accept its monopoly of military-technical expertise. This centralised, circular and assiduously compartmented system placed enormous power in the hands of the country's civilian leadership, but that leadership was undemocratic and inculcated no sense of accountability to citizens or even to executive structures outside a tightly bordered and inbred network.

The positive side of this inheritance is that Ukraine does not possess an army of intriguers, and its military leadership has kept aloof from political struggle. The negative side is that the vacuum ceded by the Party leadership has been filled by the President. According to Leonid Polyakov of the Ukrainian Centre of Economic and Political Studies (aka Razumkov Centre), the President has issued over 300 decrees on military matters. According to the Constitution, adopted on 28 June 1996, he has every right to do so. The President is Commander-in-Chief of the Armed Forces and other military formations; he directs national security and defence policy, appoints senior military commanders, establishes, reorganises and disbands executive structures and declares war, states of emergency and military mobilisation. Because he cannot possibly attend to all of these tasks himself, the Western observer would naturally assume that they fall within the day-to-day provenance of the Cabinet of Ministers and government, which is at least partially accountable to the parliament. But in fact, many of these tasks are entrusted to the President's Administration, which employs approximately 1,000 people, is not mentioned in the Constitution and is not accountable to anyone apart from the President himself. The National Security and Defence Council (NSDC), which according to Article 107 of the Constitution, is chaired by the President and 'coordinates and controls the activity of bodies of executive power in the sphere of national security and defence' has at times had an impressive staff in its own right, and two of its three Secretaries have often exerted a strong, beneficial and widely respected influence. But their ability to do so depends entirely upon their relationship with the President.

In this *schéma*, the *Verkhovna Rada* (parliament) has a decidedly subsidiary role. According to Article 85 of the Constitution, the *Rada* has the authority to approve the state budget (including defence budget), but in the real world this is very different from the ability to scrutinise expenditure. Until recently, the budget presented had only nine categories (though a measure of progress since serious reform began is that it now has 30). Under the same article (paragraph 22), it also has the power to 'confirm' the numerical strength of 'the Armed Forces, the Security Service of Ukraine and other military formations', though in practice the numerical strengths and even the budgets of many of these other force structures are secret. The *Rada* also has the power to enact laws, and in the defence and security sphere it has enacted several highly detailed ones. Yet the influence of these laws on the actual programmes that define defence policy is limited, and the *Rada*'s input into these programmes (which have binding force but do not have the status of 'laws') has been virtually nil. The Programme of Armed Forces Reform and Development 2001–5 has been published—which is an immense step forward—but it was approved by the President, not the Parliament. Finally, the *Rada* has a Standing Commission on Security and Defence, but unlike committees of the US Congress and the House of Commons Defence Committee, the Commission does not have a standing corps of expert advisers. Whereas some members of the Commission are active, well-informed individuals who take their responsibilities seriously, this is not true of everyone, and it could not be said that the Commission has emerged as an effective focal point of opposition to the status quo. In sum, whatever the formal powers of the *Rada*, it has limited capacity. In contrast, the British Parliament and the House of Commons Defence Committee have relatively few powers in the defence sphere but a considerable influence thanks to a far greater contrast: the presence in the United Kingdom of an independent and critical mass media, a large, confident and active community of experts and an organised and articulate opposition which is able to draw upon these resources.

On the margins, the weakness of civilian democratic control in Ukraine is reinforced by a linguistic confusion, which Ukraine's NATO partners have unwittingly abetted. In the Ukrainian and Russian languages, the dictionary definition of the term 'control' (translated as *kontrol'*) is 'checking' or 'verifying', which in political-administrative terms connotes 'oversight' at most. Only when Ukrainians are exposed to Western practice do they grasp that our concept intrudes into other domains: *upravlinya* (direction), *zaviduvannya* (management), *keryivinstvo* (administration), *naglyad* (supervision), as well as *kontrol'*.

Until recently, this state of affairs suited the Armed Forces perfectly. For the majority of its commanders, and those of other force structures, civilian democratic control means control by an elected president, and that is exactly what Ukraine has. Yet thanks to the dynamics of defence reform—and, alongside it, cooperation with NATO—a growing number have come to appreciate that in a democracy (however limited) with a market economy (however distorted), defence policy will suffer if it is not the outcome of broad deliberation between executive and elected institutions: institutions that will be called upon to finance it and justify it. For the same reason, a growing number of people now understand that those who work in these institutions had better be knowledgeable than not.

Possibly, the most outstanding and influential example of this is Ukraine's current (and sixth) Minister of Defence, Yevhen Marchuk. Unlike all but one of his predecessors, Marchuk is not a military man, although he has long held the pro forma rank of Army General of the Security Services (SBU). Alongside his experience in the SBU, which he headed from November 1991 to July 1994, he has very broad inter-agency experience and was both a Deputy Prime Minister and Prime Minister of Ukraine. Marchuk also understands the two strategic sectors of Ukraine's national economy, energy and defence industry, and he understands the inner workings of the trans-national networks that dominate both. For this reason, he is well equipped to deal with the commercial networks and 'parasitical structures' that have grown up inside and in association with the defence establishment. For the same reason, he is able to talk to the heads of civilian ministries with authority and at least on equal terms. This is something that a military minister, particularly a Soviet trained military minister, is simply unable to do. There are lessons to be learned here, and it would be surprising if some of Ukraine's military professionals were not learning them.

Yet whatever lessons are learned in the defence sector, they will have only a limited impact on Ukraine's democracy and security unless they extend to the security sector. But for historical, political and institutional reasons, the learning curve is proving to be even less smooth there than it has been in the Armed Forces. Again, the historical factors are highly relevant. Following the sanguinary, formative period of the Soviet state, the Soviet Armed Forces lost most of their internal functions. These were for the most part entrusted to substantial, highly militarised formations of the security services (the *VeCheka* and its successors) and the Ministry of Internal Affairs. The practice extended to guarding borders, whose first line of defence became State Border Troops, which by the 1980s had evolved into a heavily militarised force of 400,000 subordinated to the KGB. This separation of the Armed Forces from the unpalatable and deeply politicised tasks of suppressing internal conflict and confronting their own citizens became part of its ethos, and it has contributed to the respect with which they were held by much of society. This tradition continues to be observed in Ukraine, and its positive effects continue to be felt. In Ukraine, only 11.9% of the population express trust in the *militsia* (civil police), 11.8% trust the courts and (before the tape scandals damaged their reputation further) 20.2% in the Security Services (SBU). Yet 55% have either 'complete trust' in the Armed Forces or are 'inclined to trust' them.

Yet this strict distinction between external and internal functions has a number of negative sides. For one thing, it reflects the inbred Leninist instinct of divide and rule and the fear of what might occur if the Armed Forces had a monopoly of force. This instinct and practice continue to flourish in post-Soviet Russia and Ukraine. Indeed, in both countries the proliferation of force structures has continued since the Soviet collapse, and it is even given legitimacy by Ukraine's Constitution which (Article 17) speaks of 'other military formations' alongside the Armed Forces—and maintains the Soviet distinction between the term 'military' (which can apply to any subordination) and the term 'Armed Forces' (often capitalised as *Zbroynyi Syly*) which can only be used to apply to armed services subordinated to the Ministry of Defence. Therefore, schemes of civilian democratic control confined to the Armed Forces will be dangerously incomplete and will do very little to diminish the gap between state and society.

All of this needs to be borne in mind by Westerners who expect the *army* in Ukraine to be remodelled into an institution capable of addressing soft security challenges. To some extent, it must be, it can be—and, to some extent, as we note below (pp. 162–3) that is what is taking place. Yet in the Ukrainian system, most of this burden is meant to fall upon other shoulders, and in principle there is nothing wrong with this, so long as those who perform these functions are well trained, well controlled and have no confusion about what state, what people and what scheme of values they are protecting. Ukrainian military professionals believe it is their primary task to provide armed defence of the country against external opponents. Although the focus today has shifted from general war to local and limited conflicts, the view that interstate wars have become a thing of the past and that current and future security threats will be transnational, internal or 'soft' would be disputed with conviction even by many of the most reformist Ukrainian military officers, all of whom could cite conflicts in the former Soviet Union, the Balkans and Iraq in support of their views. The fact that a strong element of corporate interest is present in these perspectives does not detract from their force or the conviction with which they are held.

Defence Reform

In Ukraine progress in civil–military relations has lagged well behind progress in defence reform. This is a telling contrast with recent members of NATO, most of whom were relatively swift in democratising the defence sphere, but often lacklustre and amateurish in addressing the conceptual, material and operational dimensions of defence reform.[4] Although Ukraine has significant accomplishments to its credit in this area, here too, progress needs to be measured against a Soviet rather than a NATO template.

Progress has come in two waves: post-independence, when it was most dramatic, and since December 1999, when it has been more methodical. In the post-independence period, the fact that troops of the Soviet Armed Forces, MVD and KGB numbering 1.4 million men were substantially reduced and thoroughly resubordinated—all of this without conflict and upheaval—was a contribution to European security second only to the country's unilateral nuclear disarmament. But it was an early and finite contribution, not an ongoing and dynamic one. Fortunately, two subsequent contributions have provided such a dynamic.

One of these has been the NATO–Ukraine relationship, which we survey in part two. The second dynamic was that launched by Ukraine's adoption of a radical, but coherent set of first principles about the country's security interests, its likely security threats and the type of military formations required to maintain security. Ukraine's first National Security Concept (1997) drafted by the analytical staff of the National Security and Defence Council under the stewardship of its Secretary, Volodymyr Horbulin, was a model statement of first principles. It assaulted the general war ethos by stipulating that in conditions where both state and society were weak, the prime security challenge would be to forestall and resolve local crises, emergencies and conflicts and prevent them from being exploited by actors (internal and foreign) with ulterior political ends. Proceeding from this analysis, the Concept identified 'the strengthening of civil society' as the first of nine security challenges for Ukraine.

However, it has taken a considerable period of time to translate these concepts into practical programmes. In the first official stage of armed forces development (1991–6), Ukraine established the legislative basis, as well as the institutional and command structures for independent armed forces. It repatriated over 12,000 officers and warrant officers who refused to take an oath of allegiance to Ukraine (and absorbed 33,000 military servicemen from other parts of the USSR). It demobilised over 300,000 servicemen. It also disarmed the world's third largest nuclear force, removing the last nuclear warhead from its territory by 1 June 1996. Yet at the start of the second stage (1996–2000), Ukraine's armed forces still were a bloated, grossly underfinanced establishment of 400,000, lacking an authoritative, coherent and realistic scheme of transformation and development. During this second stage (in 1997), Ukraine also adopted a State Programme of Armed Forces Development to 2005. This programme had its merits, but it did nothing to assault the general war ethos, which pervaded the military establishment and did not show much recognition of the country's severe economic constraints.

Not until the adoption of the State Programme of Armed Forces Reform and Development 2001–5 were steps taken to give these principles definite content. Imprecise and unrealistic as some of the Programme's aims and targets have been, they have been steadily revised under pressure of expert criticism and unforgiving economic reality. Between 2000 and 2002, the State Programme was supplemented by others, notably the Concept of the Armed Forces 2010 and the State Programme of Armed Forces Transition Towards Manning on a Contract Basis, designed to transform the 288,600 mixed conscript-volunteer force (as of 1 January 2003) into a much smaller all-volunteer force by 2015.[5]

In operational terms, the Programme mandated the establishment of a structure of command, operations and logistics that was truly joint and based upon the three Operational Commands (OC) (Northern, Western and Southern) originally established in 1998.[6] Over the five year period, they were to be transformed into structures capable of mobilising, commanding and supporting 'multi-component' forces in the tasks of responding to peacetime emergency, as well as preventing, containing and 'neutralising' armed conflict. In structural terms, the Plan called for a reorganisation of the Armed Forces into three components: Forward Defence Forces (comprising Strategic Conventional Deterrent Forces, Rapid Reaction Forces and Covering Forces), Main Defence Forces and Strategic Reserve Forces. As in the UK, this joint approach has not been intended to eliminate separate armed services (Ground Forces, Air Defence Forces, Air Forces and Navy) but provide a radically different approach to conducting operations. Although the Soviet Armed Forces, too, had a combined arms structure, it only came into play above the tactical level. In Ukraine the combined arms focus now extends to sub-tactical and low intensity activity. This shift in emphasis is now reflected in training and education. Above officer commissioning level, education is now joint service and also includes non-MOD force structures.

Whilst very positive above this programme, NATO members of the JWGDR were sceptical about whether a number of its goals could be achieved within current institutional and resource constraints. For example, the General Staff and subordinate to them, the OC Commander will have authority over components of other force structures (e.g. Interior Troops and Emergency Situation Troops) in a conflict or emergency situation. But what mechanisms are being provided to create a common operational culture and a common basis of training in peacetime? Apart from points of scepticism, there were also serious points of criticism, not least with regard to the extremely modest force reductions mandated: from 310,000 servicemen and 90,000 civilians in January 2001 to 295,000 servicemen and 80,000 civilians in 2005. For all the merits in the Programme, it therefore did little to persuade the wider NATO community that reform would bring resources and capabilities into balance.

Today, this would be a less justified conclusion to draw. Under Army General Volodymyr Shkidchenko (who replaced Army General Oleksandr Kuzmuk on 24 October 2001) several of NATO's concerns were quietly addressed. Shkidchenkos's successor, Yevhen Marchuk (who replaced him on 25 July 2003) has gone further and with greater boldness. Despite its focus on local war and low-intensity conflict, the State Programme envisaged equipment holdings vastly disproportionate to these requirements: 3,276 tanks, 4,203 AFV, 3,684 artillery pieces and 406 fixed-wing combat aircraft. In January 2002, Shkidchenko's revised these figures downward by more than 30% (to a maximum of 2,000 tanks, 3,500 AFV, 2,000 artillery pieces and 300 combat aircraft). To be fair to Shkidchenko as well as his predecessor, Kuzmuk, these totals reflected an estimate of how much could be sold on the international market, and both ministers envisaged further reductions over time. Shkidchenko also put in place a well-planned scheme of base closures. A total of 43 military bases were closed in 2001. This expanded to 69 in 2002, and (as of October 2003) the total was projected to rise to 120.[7] Building on Shkidchenko's plans, Marchuk authorised an establishment of 160,000 servicemen (and

40,000 civilians) in the Armed Forces by the end of 2005 (a reduction of 175,000 from the figures presented in the State Programme). As Secretary of the National Security and Defence Council, Marchuk was also able to secure presidential and parliamentary backing for a substantial increase in the defence budget: from 3,541.2 mn *hryven* (UAH) (approximately $680 mn) in 2001, to UAH 4,282.7 mn (approximately $823 mn). Ukraine's harshest knowledgeable critic of its defence reform, Anatoliy Grytsenko (now President of the Razumkov Centre) concluded in 1999 that Ukraine's Armed Forces should not fall below 150,000 troops and that defence spending would need to rise to 2.5% of GDP. Although defence spending is still no more than 1.7% of GDP, projected totals will now fall very close to Grytsenko's minimum figure. In several spheres, training and combat readiness have also improved substantially. In 2003, Air Force pilots in the Rapid Reaction Forces flew an average of 90–100 hours per year (some four times greater than the average figure for the Air Forces in 1999), and graduates of the military aviation institute flew 130 hours.[8] But in the longer term Marchuk's two greatest accomplishments are likely to stem from his assault upon 'parasitical [commercial] structures' in the military establishment—a risky and possibly perilous task—and the Defence Review, set in motion by Presidential Decree No 565 in July 2003.

The Defence Review, due for completion in July 2004, has subjected the entire organisation of defence to painstaking and remarkably transparent review, in close cooperation with the Ministry of Finance and experts in NATO, with whom all relevant data has been shared. Whilst not designed to overturn existing plans, it will certainly refine them and define the basis upon which they can be elaborated and implemented. For example, the State Programme called for 12 out of 13 divisions of the Ground Forces to be transformed into brigades. But without a rigorous assessment of inventories, infrastructure, personnel and costs, there is no definite way of knowing whether this scheme is the right one, let alone what its precise costs and collateral effects will be. The Review will revise plans and create a sound basis for further revision. In the view of some NATO experts participating in the process, it will approximate the standards of informativeness, transparency and realism found in the defence reviews of NATO member states.

The Review will also strengthen realism regarding professionalisation, a subject that generates mixed feelings in Ukraine and confusion in the West. Given the historical and economic factors that bedevil professionalisation in the post-Soviet context, the emphasis placed upon it in Ukraine is noteworthy. Although elements in the military establishment have, with reluctance, accepted this policy, formalised by the State Programme of Armed Forces Transition Towards Manning on a Contract Basis (2001), there has been no resistance to it. The number of professional ('contract') servicemen has risen steadily from 29,446 in January 1999 to 42,300 in October 2003, which represents 26% of non-commissioned ranks. The aspiration (plan) is for this total to reach 50,000 by January 2005 and 70,000 by 2010, perhaps as much as 72% of other ranks.

The reluctance that still exists in some quarters is, once again, the product of the Soviet legacy. The Soviet Armed Forces was a conscript force not only in fact but in ethos. To the Soviet officer, conscription is what gave substance to the principle of 'unity of Party, army and people'. For many senior Ukrainian officers, conscription is still seen as an institution tying the Armed Forces to ordinary people and making citizenship a meaningful term. Like their equivalents in the Russian Federation, many instinctively equate professionals with mercenaries and believe that there is something deeply immoral in 'hiring' soldiers to defend the security of their own country. Today this perspective is not shared by most junior officers, who have learnt that conscription detracts from their own professionalism. First, they can see that in Ukraine as in Russia, the best and the brightest have the connections to evade military service. (According to one 2000 report, 'some 90 per cent of conscripts are either released from duty or enjoy postponement rights').[9] Second, the low quality of the conscript cohort that actually serves obliges many junior officers to perform tasks that in the British and American armies are performed by NCOs. To be sure, a long-term, technically skilled professional NCO corps on NATO lines, which is slowly coming into being in Ukraine, will gradually alleviate this burden, but the poor quality of conscripts pulls the whole system down. Society, too, does not share the views held by traditionalists. In a January 2002 poll, 33% of respondents identified the professional soldier as a 'staunch defender of the native land and true patriot', 21.5% identified him as 'a fighter who is able to perform the most complex tasks', 12.5% described him as 'the present-day serviceman, but better maintained', and only 11.8% equated him with 'a mercenary fighting for money'.[10] Both Ukraine's policy and its implementation demonstrate that the traditional view has not been upheld.

This is less remarkable than the fact that resources are being found for professionalisation and its corollary, force reductions. The budgeted costs for conscription are extremely low. In 2001 the budgetary cost of the conscript cohort was $6.5 million, roughly one-and-one-half per cent of the defence budget. Hidden costs (food, housing and other support) drive these costs up considerably, but they are still low. As opposed to the conscript wage of $50 per annum, the salaries of *kontraktniki* (contract soldiers) are $50 per month. Reductions are also expensive, because if they are undertaken properly, they require base closures, and if they are undertaken legally, they demand the retraining and resettling of officers, who can only be dismissed if they have civilian employment. Whilst the base closures are underway, resettlement is an enormous institutional and financial burden. Hence, there is the worrying possibility that the officer corps (already considerably disproportionate to the size of other ranks by NATO standards) will become a still larger proportion of the army as reductions are carried out. The equally worrying possibility is that professionalisation will not extend far beyond professionally attractive wages, whereas the real challenge is to provide an infrastructure for the continuous and progressive training regime and superior services enjoyed by NATO professionals, who have the skills and long-term motivation required by a technologically advanced force.

Even in summer 2003, the dynamics of modernisation, stagnation and decay were precariously balanced in Ukraine's Armed Forces. Today they are balanced rather more favourably and less dangerously. In 2002 Major General Valeriy Muntiyan, Assistant to the Defence Minister for Budget and Financial-Economic Activity, could state that without a radical revision of financial support, 'the Armed Forces have no more than five years until self-ruination'. That forecast should still be borne in mind if today's momentum is not maintained. As we have already noted, there is nothing in the political or geopolitical context which provides a guarantee that it will be.

International and Transnational Influences

Ukrainians are very conscious of the fact that the country's strategic orientation has rarely depended upon its wishes. The political, military and security establishments of Ukraine proceed from the assumption that the country's geopolitical vulnerabilities will, over the short, medium and long-term, act as a powerful constraint on 'the art of the possible'. As already noted, Ukraine is in a vulnerable geographical position between the Black Sea and seven neighbours. It is also a rear area of the Caucasus and the Balkans, and it has been subject to external pressure in both the Balkan and Chechen conflicts.[11] Moreover, its most influential and closest neighbour in terms of cultural affinity, the Russian Federation, is reconciled to Ukraine's independence *de jure* more than it is *de facto*. Until 2003, the Russian Federation officially refused to accept that the border between the two states should be demarcated; many of its official representatives speak of Ukraine as an 'ally' (thus refusing to recognise the country's non-aligned status), and Russian ministers have on several occasions referred to Ukraine's neighbour, Moldova, as a state bordering Russia. All of this, combined with a number of Russian actions (e.g. the use of Crimean bases to train troops for combat duty in Chechnya) creates the impression that respect for Ukraine's sovereignty, endlessly reiterated *pro forma*, does not fully conform to real Russian thinking and practice. More fundamentally, the Russian authorities see no contradiction between independence and 'integration'. The Treaty on the Single Economic Space, which President Kuchma signed (over the public objections of several of his own ministers) in response to Russian pressure and perceived EU coolness, establishes a series of undertakings contrary to Ukraine's long-standing official goal of joining the European Union, and the EU has stated as much itself. The basing of the Black Sea Fleet, as well as its air, intelligence and Naval Infantry components in Crimea (until at least 2017) adds to other concerns that Ukraine could be involuntarily drawn into conflict with third parties.

A clear majority of Ukrainian citizens believes that non-alignment is indispensable to good relations with neighbours, as well as political stability. Within the political establishment, an influential number of avowed 'centrists' have gone two steps further: favouring a large measure of economic integration with CIS countries and security integration with the West. Since Leonid Kuchma was elected President in July 1994, Ukraine has therefore pursued a 'multi-vector policy', with shifting degrees of emphasis accorded to each vector in response to internal and international circumstances. In this context, the NATO–Ukraine relationship—put on a solid footing when Ukraine joined Partnership for Peace (PfP) in February 1994 and enhanced by the signing of a Charter on a NATO–Ukraine Distinctive Partnership in July 1997—has had a unique importance.[12] Despite the Russian factor, the relationship has continued to be firmly supported by President Kuchma who, like most centrists and much of the left, do not wish to see the de facto primary vector, Russia, become the sole vector influencing Ukraine's choices and development.

Despite growing Russian influence and the deep controversy aroused by NATO's intervention in Kosovo, the NATO–Ukraine relationship has taken on a more intense and practical form since President Kuchma decreed the start of the latest defence reforms in December 1999. If prior to 1999, it was first and foremost a political relationship, cemented by a web of military-to-military contacts, it is now at least as much a military-technical relationship. Whilst today perhaps even more than a few years ago, NATO is seen as an essential counterweight to Russia, today there are far fewer hopes that NATO can serve as Ukraine's primary vehicle for 'entering Europe', as the role and requirements of the EU are well understood, even if they are understood in sullen and resentful terms. Whatever the NATO–Ukraine relationship brings in geopolitical terms, its day to day business, in the words of former Minister of Defence, Army General Oleksandr Kuzmuk, is 'supporting defence reform in the country'.[13]

Although a majority of influential players see Russia as the prime partner in defence industrial collaboration—and here perhaps more by necessity than by choice—no one of significance in Ukraine's military establishment believes that Russia should be the partner of choice in carrying out defence reform.[14] First, in all the areas where progress is sought—low-intensity operations, joint operations, professionalisation, planning and budgetary transparency, civil–military collaboration—NATO is seen as the repository of experience and expertise. In contrast, Russia's inconsistent and internally contested reforms offer a poor model, and the performance of Russian combat forces in Chechnya does not lend itself to imitation. Second, Russia's aims are mistrusted and its methods regarded with suspicion. As noted by three specialists in *Zerkalo Nedeli*:

> An indicative example [of the wrong] approach is provided by the Agreement between Ukraine and Russian concerning the conditions

regulating the basing of the Russian Federation Black Sea Fleet on Ukraine's territory. Up to now, neither the Ukrainian public nor the parliamentarians of Ukraine have seen the full text and conditions of this extremely important document for Ukraine and its future.[15]

A different but complementary point is made by Leonid Polyakov, Director of Military Programmes of the Razumkov Centre:

So far, Russian officials, unlike NATO's, have never voiced their concern about the weakness of Ukraine's defence or the slow pace of its military reform. One might infer that Ukraine's problems in building its Armed Forces are simply more acceptable to Moscow than Ukraine's success in that area.[16]

In contrast, it is obvious that NATO seeks to develop Ukraine's institutional capacity, rather than undermine it, and it is also obvious that NATO is less keen on having Ukraine as a member than Ukraine is itself.

NATO's role is institutionalised by several mechanisms unique to the NATO–Ukraine relationship. The principal mechanism, the NATO–Ukraine Joint Working Group on Defence Reform, was established under the Distinctive Partnership, but energised by the State Programme. Within this framework, cooperation has now moved beyond the formal exchange of ideas to a scheduled process of audit and consultation. In 2001 Ukraine became a fully active participant in NATO's Planning and Review Process (PARP): a PfP programme requiring each participating country at regular intervals to supply NATO with a detailed inventory of its military assets and, jointly with NATO, identify real costs, as well as capabilities in short supply or surplus to needs. Several additional mechanisms (e.g., the NATO Liaison Office in Kyiv) have also been devised to facilitate cooperation.

Since the 23 May 2002 declaration (defining NATO membership as the 'ultimate' goal of Euro-Atlantic integration), the relationship has taken an additional step forward. This declaration represented a radical change of policy, for up to that point policy had been predicated on non-alignment and, hence, a firm (but not always clear) distinction between 'integration' with NATO and membership of it. Because of its 'long-term' focus, the NSDC's declaration was not accompanied by an official Ukrainian application to join NATO. Because the declaration emerged at a time when President Kuchma's relations with most NATO governments had reached an impasse—and because NATO makes decisions by consensus—NATO did not issue Ukraine with an invitation to submit a Membership Action Plan. Yet both sides grasped that the moment had to be captured and exploited. The NATO–Ukraine Action Plan was the result.

As an officially Ukrainian document, drawn up in consultation with NATO and approved by the NATO Council in Prague (22 November 2002), the Action

Plan represents the definitive statement by both parties about what Ukraine must do to achieve its goal of 'full Euro-Atlantic integration'. It is an ambitious document with a clear message: it is the country, not the army that joins NATO, and the state of the country will be the principal benchmark that NATO will use to assess Ukraine's progress. This should be sobering and possibly unwelcome news to those tempted to use cooperation (e.g., the dispatch of the 1,600 strong Ukrainian military contingent to Iraq) as a substitute for in-depth internal change. Valued as this cooperation is, here, too, the message is clear: in itself, political–military cooperation will merely prolong the status quo, under which the 9 July 1997 Charter on a Distinctive Partnership, rather than membership, 'remains the basic foundation of the NATO–Ukraine relationship'. However the political dimension of the Action Plan might be interpreted by President Kuchma and his key allies, the military dimension is being addressed with due seriousness, as are the mechanisms set out for NATO consultation and review, notably Annual Target Plans, designed to establish a scheduled and measurable sequence for the Action Plan's implementation.

In today's post-9/11 world, four politico-military dilemmas are felt acutely inside Ukraine. First, the focus of the United States and other key NATO players on the 'war on terror', the war in Iraq—and the buffeting that trans-Atlantic relations have suffered as a result of the two—almost guarantees that Ukraine will occupy a diminished role in their scale of priorities. But by how much? There is surely no comfort to be derived from the possibility that one of the corollaries of the war on terror—greater interest in energy partnership with Russia—might diminish the inclination of the West to challenge Russia's increasingly assertive claim that the CIS is its 'zone of special interests'. In response to these anxieties, the natural impulse is to demonstrate Ukraine's usefulness both in the war on terror and the war in Iraq, as demonstrated by the decision to deploy the Coalition's fourth largest military contingent to that country.

This gives rise to the second dilemma. This step is highly unpopular in Ukraine. Until 9/11 Ukraine faced no significant threat from terrorism. Indeed, the relative absence of such a threat—which has plagued Russia since the start of the second Chechen war at least—has reinforced wishes to live in a 'normal' country and neither return to Russia's fold nor share its burdens. Yet a growing number of people are now asking whether Ukraine's complicity in US policy is exposing the country to the very threats that it has long avoided. Thus, many radical democrats (e.g., Yulia Tymoshenko) can agree with Communists that Ukraine's love affair with NATO is unrequited, that it has put lives on the line for interests that are not its own, and that Ukraine has received nothing tangible in return.

This dilemma has an affinity with a third. NATO is a collective defence organisation. Ukraine has based its policy and planning on the assumption that it might—and most probably will—be obliged to defend itself. National armed forces 'closely resembling Euro-Atlantic standards and practice' were already a stated aim of the State Programme 2001–5, yet these forces were designed to act independently and, if necessary, alone. Even before the adoption of this programme, Ukraine made a noteworthy if finite contribution to collective defence by assigning contingents of forces to NATO-led peace support operations—indeed by 2003 over 20,000 Ukrainian servicemen had served in peace support operations under the aegis of NATO or the UN. Yet this did not diminish the emphasis on self-reliance. To what extent should this orientation change in the absence of hard assurances that the Action Plan will be followed by Membership Action Plan and, in the foreseeable future, membership? Will even membership provide a substitute for national defence? The fact that NATO sees little contradiction between national defence and collective defence may assuage anxieties, but will not quell them.

The fourth dilemma stems from the fact that Ukraine's defence programmes were drawn up in the pre-9/11 world and prior to the latest, more ambitious phase of Russian policy. In October 2003, the Russian Federation set out unilaterally to resolve a long-standing border dispute in the Sea of Azov by constructing a causeway under the protection of Ministry of Emergency Situation troops in the Kerch Strait—without, incidentally, incurring any public reproach on the part of NATO.[17] This action, and the lack of a NATO response, has had a deeply unsettling effect in Ukraine. Since 2000, the justification put forward by Ukraine's MOD for force reductions has been twofold: economy and the diminished threat. Is it now necessary to revise these assumptions?[18]

The calm and considered answer should be no—and for two reasons. First, it is wrong to equate economy with impoverishment and force reductions with a less capable force (as many Ukrainians still do). Both NATO and at least two Ukrainian Ministers of Defence have regarded modernisation, quality and force reductions as a seamless whole. What would rescue a force of 288,000 other than an economic miracle in the country?

Moreover, in what way does the Tuzla crisis refute long-standing threat perceptions? From at least 1997, these threat perceptions have emphasised the 'combination of factors' linking Ukraine's internal weaknesses and its vulnerable geopolitical position. Although total war has been removed from the official lexicon, official thinking has not minimised the likelihood of local war, let alone the potential for relatively small conflicts to escalate in intensity and in geographical scale. As Marchuk's predecessor, Army General Volodymyr Shkidchenko stated in 2002:

> the probability of a large-scale and prolonged war is low. Ukraine has no enemies to wage a total war, and ... one should not expect the appearance of such enemies in future.

172 *Civil-Military Relations in Postcommunist Europe*

But he preceded this statement by noting that 'transient, limited, possibly very fierce local interstate conflicts' remained possible. As long ago as 1997, the General Staff determined that the role of Armed Forces in this schéma would be to 'set up a zone which would make it possible to direct or influence the processes occurring outside it': a formula certainly applying to 'processes' in a neighbouring country. How is this schéma in any way irrelevant to the Tuzla crisis? It would seem to be tailor made for it.

Ukraine is accustomed to complexity, and these are complex dilemmas. But that is not to say they will be resolved to Ukraine's advantage, or even NATO's, if they are not treated with respect and understanding by Ukraine's partners.

Notes

[1] The views expressed in this article are the author's and not necessarily those of the UK Ministry of Defence.

[2] According to a poll of the Ukrainian Centre for Economic and Political Studies, 40.4% described information about defence reform as 'insufficient', 39.6% believed that 'we are actually barred from trustworthy information', and only 4.4% believed that a 'planned process of reform' was taking place. Andriy Bychenko, 'Ukrainian Citizens about their Army', *National Security and Defence* (Kyiv) 1 (2002), p. 40.

[3] In the same January 2002 UCEPS poll, respondents placed 'military policy' as the country's eighth national priority (2.4%), well behind the first priority, 'economic policy' (32.9%). Mykola Sungurovskiy, 'Military Reform: Progress Against a Background of Stagnation', in *National Security and Defence*, p. 59.

[4] See for example, the author's study, 'NATO's New Members: A Lesson for Ukraine? The Example of Hungary', *CSRC G86*, September 2000.

[5] The total number including civilian personnel stood at 382,200 in January 2003, down from 405,200 in January 2002 and 415,800 in January 2001.

[6] Unlike the Military District in the USSR, which was a territorial-administrative formation without command responsibilities, the Operational Command is 'a permanent operational and strategic formation assigned operational and mobilisation missions ... responsible to defend territory and provide logistic and other support to forces in its sector regardless of their subordination'—and do so not only in war but 'conflicts of various intensity'.

[7] Ukraine's MOD defines a base as 'a military object (unit, centre, etc.) with supporting infrastructure'. It would appear that here, as in general Ukrainian and Russian usage, the term 'object' (ob"ekt) corresponds to the English language term 'facility'.

[8] Yevhen Marchuk interview with BBC World Service Ukrainian Section, 9 December 2003, 'Challenges and Answers: Why Ukraine Needs Military Reform'.

[9] *National Security and Defence*, 1 (2000), pp. 10–12.

[10] Andriy Bychenko & Mykhailo Pashkov, 'A Professional Army in Ukraine: The Views and Assessments of the Population', *National Security and Defence*, 5 (2002), p. 32.

[11] Russia's dispatch of a 'humanitarian' convoy to Yugoslavia (halted on the Hungary–Ukraine border) in April 1999, its redeployment of the intelligence ship *Liman* (and initial preparation to redeploy other vessels) from Sevastopol to the Adriatic and its plans to transit Ukraine with Airborne Troop reinforcements after the 'brilliant dash to Pristina' in June 1999 provoked anxiety and, in some quarters, alarm. For a more comprehensive discussion, see James Sherr & Steven Main, *Russian and Ukrainian Perceptions of Events in Yugoslavia*, Paper F64, May 1999, pp. 2, 17–24 (Conflict Studies Research Centre, RMA Sandhurst, Camberley).

[12] More than 500 bilateral activities are scheduled between NATO Allies and Ukraine in 2001, as well as 250 multilateral activities with NATO.

[13] Statement to the NATO–Ukraine Joint Working Group on Defence Reform, October 2000.
[14] The number of Ukraine–Russia activities has gradually risen every year from 28 in 1998 to 52 in 2000. But this puts it on a par with Poland and at about 60% of the Ukraine–UK level.
[15] Eduard Lisitsyn, Rustem Dzanguzhin, Aleksandr Goncharenko, 'Civilian Control and the System of National Security of Ukraine' [*Grazhdanskiy kontrol' i sistema national'noy bezopasnosti ukrainiy*], *Zerkalo Nedeli*, 35: 410, 14–21 September 2002.
[16] Leonid Polyakov, *National Security and Defence*, No 12 2000, p. 15.
[17] At the height of this crisis, Russia's Prime Minister, Mikhail Kasyanov spoke of the necessity to remove (*ubrat'*) Ukraine's border troops and the Chief of the President's Administration went so far as to state that if Ukraine resisted, Russia should 'drop a bomb'.
[18] At least one respected commentator has called for a reconsideration of the fundamentals of defence policy and seriously 'contemplate forceful resistance to a state whose military potential significantly exceeds that of Ukraine'. Valentyn Badrak, 'The Right to Use Force' [*Pravo na silu*] in *Zerkalo Nedeli* [Mirror of the Week], 10 November, 2003.

Index

Academy for Non-Commissioned Officers, Latvia 58
Afghanistan 8
  Croatian troops 84
  Polish troops 46
  Romanian forces 107, 108
  Russian intelligence 148
Allied Rapid Reaction Corps (ARRC), Poland 42
Antall government 23
Antunović, Zeljka 88
Armed Forces of the Republic of Croatia (OSRH) 74, 77, 79, 82, 84–7, 88, 89
Armed Forces of Serbia-Montenegro (VSCG) 4, 120, 122, 123, 124, 125–8, 130, 131
Association of Veterans and Invalids of the Homeland War (HVIDRA) 82, 87
Atlantic Council of Serbia and Montenegro 124
Audit Court comptroller, Romania 100
Audit Directorate, Romania 101
Audit Office, Poland 37
Auftragstaktik 58
authoritarianism 5, 12, 14, 158

Baltic Air Surveillance Network (BaltNet) 56
Baltic Defence College 56, 58, 61, 62
Baltic Naval Squadron (BaltRon) 56
Baltic Peacekeeping Battalion (BaltBat) 56, 61, 66
Baltic Security Assistance Group of States (BALTSEA) 61
Baluyevsky, General Yuri 151
Basescu, President Traian 107
Beara, Ljubiša 130
Belarus 5
Bellamy, Alex J. 4, 10, 71–93
Bezborodov, Nikolay 140
Biuro Bezpieczeństwa Narodowego (BBN) 36

Black Sea Naval Task Group (BLACKSEAFOR) 107
Bokros austerity package 24
Border Guards, Latvia 54
Bosnia Serb Army (VRS) 129, 130
Bosnia-Hercegovina 8
  Croats/Muslim aggression 72
  Muslim attack 81
Buczyński, Józef 43
budget
  Croatia 78, 79, 87, 89
  Hungary 27–8, 29, 30
  Latvia 59–60, 61, 65
  Poland 47
  Romania 99–101
  Russia 139, 142, 147, 152
  Serbia-Montenegro 123, 126
  Ukraine 159–60, 165, 166
  US 65
Bush, President George 148

Ceausescu, Nicolae 96, 104
Central European Nations' Cooperation Initiative (CENCOOP) 107
Centre for Civil-military Relations (CCMR) 124, 125
Chechyna 141, 148, 149, 151, 168, 170
Chelaru, Mircea 105
Citizens' Ombudsman, Poland 37
Civic Alliance 102
civilian control 5, 12, 14
  Croatia 75, 80–1
  Hungary 20, 21–2
  Latvia 52
  Romania 95
  Serbia-Montenegro 116, 118–21
  Ukraine 160, 162
Clark, General Wesley 57
Clemmesen, Brigadier General Michael 62
Committee of Defence Affairs, Poland 36

Commonwealth of Independent States
    (CIS) 146, 148, 168, 170
Concept of the Armed Forces, Ukraine 163
conscription 1, 9, 10
    Croatia 77, 87
    Hungary 25, 26, 28
    Latvia 55, 56, 57
    Poland 43–4, 47
    Romania 103–4
    Russia 143, 144, 146
    Serbia-Montenegro 127
    Ukraine 166
Consiliul Suprem Aparare de Tara (CSAT),
    Romania 96–7, 98, 99
Constantinescu, Emil 97, 99, 101, 102, 105
Constitution
    Hungary 19–20
    Latvia 52
    Poland 34–6, 37–8
    Romania 104
    Serbia-Montenegro 122
    Ukraine 158, 159–60, 161–2
Constitutional Tribunal, Poland 37
corruption 6–7
    Hungary 17–18
    Latvia 59
Cottey, Andrew 1–16, 38
Council of Europe
    Hungary 17
    Serbia-Montenegro 129
Council of Ministers, Poland 36
Counter Intelligence Service (KOS) 119
crime, Russian military 140
Croatia 4, 6, 8, 12, 14–15
    nation-building 10
    politicisation and politics of reform 71–93
Croatian Democratic Union (HDZ) 73, 75, 78,
    81–2, 87, 88
Czech Republic
    martial history 11
    women in the military 45

Davinić, Prvoslav 123, 124
'Decision on the Size, Composition and
    Mobilisational Deployment of the CAF' 87
dedovshchina 139
'Defence Doctrine of the Polish Republic' 38
Defence Integrated Planning Directorate
    (DIPD) 100, 101
Defence Law, Russia 150
Defence and National Security Committee
    (VONS) 81

Defence Planning Council (DPC) 101
Defence Review, Ukraine 165
Defence Strategy, Croatia 74, 75
Degeratu, Constantin 105
democracy 2–7, 12, 14
    Hungary 12, 14, 17, 19–23
    Latvia 51–70
    Poland 33–8
    Romania 97–102
    Serbia-Montenegro 121–5
Democratic Opposition of Serbia (DOS) 115,
    118–19, 120, 131
Democratic Party (DS) 118, 121
Democratic Party of Serbia (DSS) 118, 121
Department for Defence Policy and
    International Relations 99
Department for Parliamentary Relations,
    Legislative Harmonization and Public
    Relations 99
desertions, Russia 140
Dindić, Zoran 118, 121, 126
Directorate of Public Relations
    (DPR) 101–2
Donnelly, Christopher 57, 58
Dubrovnik siege 72
Dukanović, Milo 120
Dunay, Pál 4, 17–32

economics 15
    see also budget
    Croatia 79
    Hungary 17
    Romania 110–11
Edmunds, Timothy 1–16, 38, 71–93, 115–35
education
    Croatia 84, 85–7
    Latvia 58
    Russia 140
EURISC 102
European Union (EU) 9, 12–13, 14, 15
    Croatia 71, 73, 74, 80, 83, 84, 89
    Hungary 17
    Latvia 55, 60, 67
    Poland 45
    Romania 99, 112
    Serbia-Montenegro 129
    Ukraine 167

Federal Republic of Yugoslavia (FRJ)
    4, 10, 12, 115–35
Felgenhauer, Pavel 149
force projection, Romania 95–114

former Soviet Union 5, 6–7, 10, 12, 14–15
   see also individual states
former Yugoslavia 8, 14, 15
Forster, Anthony 1–16, 38
FRJ see Federal Republic of Yugoslavia
Für, Lajos 22

GDP/GNP
   Croatia 79
   Hungary 17, 27, 30
   Latvia 59, 61, 65
   Serbia-Montenegro 126
   Ukraine 165
Genschel, Major General Dietrich 66
George C. Marshall Association 98, 102
Georgia, rose revolution 14
Glenny, Misha 71
Gorbachev, President Mikhail 3, 151
Gorka, Sebestyén 28
Gow, James 127
Great Patriotic War 10
Grishin, General Yuri 139
Group for Social Dialogue 102
Grytsenko, Anatoliy 165
Gyurcsány government 21

Hadžić, Mirolsav 122
hazing practices, Russia 139, 147
HDZ see Croatian Democratic Union
Herspring, Dale R. 6–7, 9, 137–55
Higher Political Directorate (HPD) 104
Homeland War 10, 78, 82
Horbulin, Volodymyr 163
Horn government 21, 24
Human Resources Department 105
Hungarian Defence Forces (HDF) 25, 26–7, 30
Hungarian Socialist Workers' Party (HSWP) 18
Hungary 3
   half-hearted transformation 17–32
   martial history 11
   nation-building 10
   1990 Open Skies agreement 106–7
   women in the military 45

IFOR, Latvia 56, 57, 61, 67
Iliescu, Ion 97, 106
Institute for Defence Analysis (IDA) 100
institutional evolution, Poland 36–7
institutional impediments, Croatia 76–7
institutional restructuring, Hungary 19–20

Integrated Concept of National Security, Romania 98
International Criminal Tribunal for the former Yugoslavia (ICTY) 78, 82, 83, 116, 119, 128, 129, 130–1, 132
International Defence Advisory Board (IDAB), Latvia 55, 57, 59, 61, 62
International Military Education Training (IMET) programme 86
Iraq 8, 11
   Hungarian troops 27, 30
   Latvian troops 56, 61
   Polish troops 46
   Romanian forces 107–8
   Russian troops 149–50, 152
   Ukrainian troops 158, 170
ISAF 84
Ivanov, Sergey 138, 141–2, 145, 146, 147, 150, 151

JNA see Yugoslav National Army
Johnson, General Sir Garry 61
Juhász, Ferenc 22–3

Kádár era 18
Kasyanov, Mikhail 147
Keleti, György 22
KFOR
   Hungary 27, 29
   Latvia 56, 57, 61, 67
   Romania 107
Kievenaar, Major General 61
Kievenaar Study 53, 60
Komisja Obrony Narodowej 37
Komitet Obrony Kraju (KOK) 36–7
Komitet Spraw Obronnych Rady Ministrów (KSORM) 36–7
Komoyedov, Admiral Vladimir 139
Kosovo 8, 11, 129, 168
   see also KFOR
   Hungarian troops 18, 24–6, 28, 29–30
   Romanian troops 107, 109–10
Koštunica, Vojislav 118, 119, 120, 121, 129–30, 131
Krga, General Branko 120, 125–6, 128
Kuchma, President Leonid 5, 157, 167, 168, 169, 170
Kuzmuk, General Oleksandr 164, 168
Kvashnin, General Anatoliy 144, 149, 150

Latawski, Paul 4, 33–50
Latvia, democracy and defence 51–70

178 Index

Latvijas fakti 66
Law on Duty to Defend the Republic of Poland 34, 36
1995 Law on the Status of Military Personnel, Romania 104, 105
Lenin, Vladimir 138
liberal democracy, Hungary 17
Long Range Management Programme 80, 83

Mahečić, Zvonimir 75
Maior, George 100, 105, 107
Mannfred Woerner Euro-Atlantic Association 102
Marchuk, Yevhen 161, 164–5
Marshall Centre 86
Medgyessy, Péter 21, 26–7, 29
media, Romania 101–2
Membership Action Plan (MAP)
    Croatia 74, 84
    Latvia 53, 57, 60, 61–2
    Romania 109
    Ukraine 169, 171
Mesić, Stipe 77, 83
Military Career Guide 105–6
Military Professional Resources Institute (MPRI) 73, 80, 83, 85
Military Reform Plan, Russia 141
Military Services Academy 150
Military Sociological Institute 43–4
Military Strategy, Romania 98
militsia 161
Milošević, Slobodan 4, 12, 14, 78, 122, 124, 129, 132
    arrest 130
    civil-military relations 116–18
    defeat 115, 119, 125, 131
Milutinović, Milan 120
Ministry of Defence
    Croatia 73, 75, 77, 79, 80–1, 82, 83, 85, 86
    Hungary 20, 22–3, 26, 27
    Latvia 53, 54, 58, 59–60, 61–2, 65, 66
    Poland 36, 37
    Romania 96, 97, 98, 99, 100–1, 103–4, 106, 110
    Serbia-Montenegro 116, 123–4, 131
Mladić, General Ratko 130
Monterey Institute 98, 100
Moskos, Charles 11
Multinational Peacekeeping Force South Eastern Europe (MPFSEE) 107
Muntiyan, Major General Valeriy 167

Nagorno-Karabakh conflict 8
Najwyzszy Izba Kontroli 37
nation-building 10, 13, 15, 72, 77
National Assembly of the State Union, Serbia-Montenegro 123
National Defence Academy, Latvia 58
National Defence College (NDC), Romania 96–7, 98, 102, 110
National Defence Committee, Poland 36, 37
National Defence Council, Hungary 19
National Defence Law on the Romanian Armed Forces 104
National Defence Strategy
    Republic of Poland 38–9, 43
    Serbia-Montenegro 122–3
National Defence University 103
national identity 10
National Military Strategy, Croatia 74
National Public Research Center (VtsIOM) 152
National Security Concept, Ukraine 163
National Security Council, Poland 36
National Security and Defence Council (NSDC), Ukraine 159, 163, 165, 169
National Security Law, Latvia 52–3
National Security Office, Poland 36
National Security Strategy
    Croatia 74, 75
    Republic of Poland 39, 40
    Romania 98
    Serbia-Montenegro 122, 126, 127
NATO 7–8, 9, 12–13, 14, 15
    see also Membership Action Plan (MAP)
    controversies 11
    Croatia 71, 74, 79–80, 84, 88–9
    Hungary 17, 19–20, 23–6, 27, 28, 29–30
    Latvia 51, 52, 53, 55, 56–7, 59, 60–2, 65, 66–8
    Poland 38, 39, 40, 41–2, 46, 47
    Romania 98, 99, 101–2, 103, 107–10, 111–12
    Russia 148
    SEEI 107
    Serbia-Montenegro 116, 119, 129–30
    Ukraine 158, 160, 162–3, 164, 165–6, 168–71
NATO-Ukraine Action Plan 169–70, 171
NATO-Ukraine Joint Working Group on Defence Reform 169
Nemtsov, Boris 143, 147
neutrality, Poland 35
Nikolić, Tomislav 78
non-commissioned officers (NCOs)

Romania 110
    Russia 139, 147, 152
    Ukraine 166
non-governmental organisations (NGOs)
    Romania 102
    Serbia-Montenegro 124
North Atlantic Parliamentary Assembly,
    Latvia 53
nuclear weapons, Ukraine 163

OECD, Hungary 17
Ojdanić, General Dragoljub 117
Onyszkiewicz, Janusz 37
1990 Open Skies agreement 106–7
Operation Enduring Freedom 107
Operational Commands, Ukraine 163–4
Orbán, Viktor 24, 25–6, 29
Organisation for Security and Cooperation in Europe (OSCE) 73
OSRH *see* Armed Forces of the Republic of Croatia

Partnership for Peace (PfP)
    Croatia 73, 74, 86
    Hungary 23–4
    Latvia 57, 60, 61
    Romania 101, 109
    Serbia-Montenegro 116, 119, 126, 129–30, 131, 132
    Ukraine 168, 169
Pascu, Minister 105
Pavković, General Nebojša 118, 119, 120, 121, 129–30
pay
    Russia 143, 145, 147, 152
    Ukraine 166
peacekeeping 8, 9, 11, 12, 14
    Baltic States 56, 61, 66
    Croatia 84, 86, 89
    Hungary 27, 29
    Latvia 56, 57
    Romania 107–8
    Serbia-Montenegro 128
    Ukraine 171
Perišić, Momcilo 119
personnel
    Croatia 87–8
    Romania 104–6
PHARE, Croatia 86
planning
    Hungary 21
    Romania 99–101

Planning, Programming, Budgeting and Evaluation System (PPBES) 100
Planning, Programming and Budgeting System (PPBS) 60
Planning and Review Process (PARP) 169
Poland 3
    martial history 11
    nation-building 10
    transformation 33–50
politics 2–7, 12
    Croatia 71–93
    Hungary 19–23
    Poland 33–8
    Romania 104, 105
Polyakov, Leonid 159, 169
postmodern military 11–12, 13, 14
professionalisation 7–9
    Croatia 84, 88–9
    Hungary 23–8
    Latvia 54–60
    Poland 38–44, 47
    Romania 102–4
    Russia 143–7
    Serbia-Montenegro 125, 127
    Ukraine 166–7
Programme of Armed Forces Reform and Development, Ukraine 160
'Programme for Integration into the NATO and Modernisation of the Polish Armed Forces' 41
'Programme for Modernisation of the Armed Forces', Poland 42, 43
promotion boards, Romania 105
Pskov division 144–5, 151
Putin, Vladimir 5, 7, 137–55

Račan, Ivica 73–4, 75–6, 78, 79, 82, 83, 87, 88
Rada Bezpieześntwa Narodowego (RBN) 36
Radoš, Jozo 77, 79, 82–3, 85, 87
Razumkov Centre 159
Reform Fund 127
regime defence, Croatia 72–3
Regional Centre for Defence Resource Management 100
Revolution in Military Affairs 13
Robertson, Lord 19
Roma/gypsy population 110
Romania 3
    force projection 95–114
    martial history 11
    nation-building 10
    1990 Open Skies agreement 106–7

Romanian Armed Forces (RAF) 95–6, 98, 100, 102–3, 106–8, 109–11
Romanian-Hungarian Joint Peacekeeping Battalion 107
rose revolution 14
Russia 5, 6, 8–9, 14
  Latvian nationals 64, 66
  Latvian threat 63, 65
  military reform 137–55
  potential threat 7
  Romanian opposition 95
  Sea of Azov 171
  Ukraine relationship 158, 167, 168–9, 171
Rzecznik Praw Obywatelskich 37

Saeima 51, 52, 59, 68
Sanader, Ivo 73–4, 76, 78, 79, 82, 88
School for Security Sector Reform 124
Security Council, Russia 141, 142
'Security Policy and Defence Strategy of the Republic of Poland' 38
Security Services (SBU), Ukraine 161
'Security Strategy of the Republic of Poland' 38, 39
Segal, David 11
Serbia-Montenegro 4, 8, 14, 15
  army in search of a state 115–35
  Croatian threat 78
  nation-building 10
Serbian Socialist Party 115
Serbs
  Croatian independence 72
  Hungarian airspace 25
  VSCG 122
SFOR
  Hungary 27, 29
  Latvia 56, 57, 61, 67
SHAPE, Latvia 61, 67
Shaposhnikov, Boris 151
Sherr, James 6–7, 157–73
Shevtsova, Lilia 137
Shishounova, Yelena 139
Shkidchenko, General Volodymyr 164, 171–2
Shpak, Colonel General Georgiy 139, 152
SIS 82
Small Constitution, Poland 34–5, 36, 37
Smallholders Party 22
social change, Romania 110
Social Research Laboratory (PBS) 44
Socialist Federal Republic of Yugoslavia (SFRJ) 3–4, 117, 132
Socialist-Liberalism, Hungary 21–2, 24, 26–7

societal impediments, Croatia 77–8
society 9–12
  Hungary 28–9
  Latvia 62–6
  Poland 44–5
  Romania 102
South East European Brigade (SEEB-RIG) 107
South Eastern Europe Initiative (SEEI) 107
South Eastern Europe Security Cooperation Group (SEEGROUP) 107
Soviet Union 3, 10
  see also former Soviet Union
Špegelj, Martin 81
Stand-by High Readiness Brigade (SHIRBRIG) 84, 107
Standing Commission on Security and Defence, Ukraine 160
State Programme of Armed Forces Development, Ukraine 163, 164, 165
State Programme of Armed Forces Reform and Development, Ukraine 163, 171
State Programme of Armed Forces Transition Towards Manning on a Contract Basis 163, 165
Stipetić, General Petar 82, 83
Strategic Defence Review (SDR), Croatia 84
Study on NATO Enlargement 60
suicides, Russia 140
Supreme Chamber of Control, Poland 37
Supreme Defence Council (CSAT), Romania 96–7, 98, 99
Supreme Defence Council (VSO), Serbia-Montenegro 117, 119–20, 121, 122, 124, 130
Šušak, Gojka 81, 88
Szabó, János 22
Szmajdziński, Jerzy 44, 47
Szumski, General Henryk 41

Tadic, Boris 121, 123, 124, 126
'Tenets for the Programme of the Armed Forces Modernisation' (Plan 2012), Poland 41–2
territorial defence, Latvia 56
terrorism 8
  Russia 148–9
  Ukraine 170
Tisza Engineer Battalion 107
tolerance training, Romania 110
Tomić, General Aco 120, 121

total defence, Latvia 56
training
    Croatia 85–7
    Hungary 26
    IMET programme 86
    Romania 98, 100, 110–11
    Russia 139, 140–1
transitology 14
Transparency International, Corruption Perception index 18
Trapans, Jan Arveds 4, 51–70
Treaty on Single Economic Space 167
Trybunal Konstytucyny 37
Tuctman, President Franjo 4, 12, 14, 72, 73, 81, 82, 85, 88
Tus, General Anton 82
Tuzla crisis 171, 172
Twin Towers attack 148
Tymoshenko, Yulia 170

Ukraine 5, 6, 8–9, 14
    flawed democracy and geopolitical anxiety 157–73
    nation-building 10
UNFICYP 27
Union of Right Wing Forces 143, 145
United Nations (UN) 56
    Croatia 84
    Romania 101–2
    Serbia-Montenegro 129
    Stand-by High Readiness Brigade (SHIRBRIG) 84, 107
United States
    budget 65
    Croatian defence 80, 83
    IMET programme 86
    women in the military 45
US Army Latvian Assessment Document (Kievenaar Study) 53, 60

Vankovska-Cvetkovska, Biljana 81–2
Verkhovna Rada 159–60
volunteers 9
    Latvia 54, 56
    Poland 43–4, 46–7
    Romania 104
    Russia 143
    Serbia-Montenegro 127
    Ukraine 163
VSCG *see* Armed Forces of Serbia-Montenegro
VSO *see* Supreme Defence Council
Vukovar siege 72

Walesa, Lech 3
Washington Treaty 60
Watts, Larry L. 4, 95–114
White Paper on National Security and Defence, Romania 98
Williams, John Allen 11
women
    Polish military 45
    Romania military 110
    Russian military 147
World Bank 103

Yeltsin, President Boris 3, 137, 141, 144
Yugoslav Army (VJ) 4, 115, 117–19, 120, 122, 123, 125–8, 130
Yugoslav National Army (JNA) 3–4, 72, 85, 117, 123, 127–8

Zemessardze 54
Zerkalo Nedeli 168–9
Žunec, Ozren 81

For Product Safety Concerns and Information please contact our EU representative GPSR@taylorandfrancis.com
Taylor & Francis Verlag GmbH, Kaufingerstraße 24, 80331 München, Germany

www.ingramcontent.com/pod-product-compliance
Lightning Source LLC
Chambersburg PA
CBHW052125300426
44116CB00010B/1782